Sarah Spiekermann

Ethics in the Age of AI

Also of interest

Ethical Perspectives on Artificial Intelligence in Biomedical Engineering
Arne Manzeschke, Thomas Wittenberg, 2026
ISBN 978-3-11-158619-9, e-ISBN 978-3-11-158645-8

Technoscientific Research.
Methodological and Ethical Aspects
Roman Z. Morawski, 2024
ISBN 978-3-11-117980-3, e-ISBN 978-3-11-118003-8

The Dynamic Equity Framework Theory.
A Global Model for Algorithmic Fairness and Cybersecurity Policy
Babajide J. Asaju, 2026
ISBN 978-3-11-914234-2, e-ISBN 978-3-11-222214-0

Demystifying Artificial Intelligence.
Symbolic, Data-Driven, Statistical and Ethical AI
Emmanuel Gillain, 2024
ISBN 978-3-11-142567-2, e-ISBN 978-3-11-142614-3

Sarah Spiekermann

Ethics in the Age of AI

A Call for Value-Based Progress

DE GRUYTER

Author
Prof Dr Sarah Spiekermann
Vienna University of Economics and Business (WU)
Welthandelsplatz 1
1020 Vienna, Austria
sarah.spiekermann-hoff@wu.ac.at

ISBN 978-3-11-167763-7
ISBN 978-3-11-167773-6 (PDF)
ISBN 978-3-11-167830-6 (EPUB)
DOI https://doi.org/10.1515/9783111677736

Library of Congress Control Number: 2026935949

Bibliographic information published by the Deutsche Nationalbibliothek
The Deutsche Nationalbibliothek lists this publication in the Deutsche Nationalbibliografie;
detailed bibliographic data are available on the Internet at http://dnb.dnb.de.

De Gruyter and Walter de Gruyter GmbH are part of De Gruyter Brill.
www.degruyterbrill.com

Questions about General Product Safety Regulation:
productsafety@degruyterbrill.com

Cover image: amtitus/DigitalVision Vectors/Getty Images
Jan Brueghel the Elder and Peter Paul Rubens, The Garden of
Eden with the Fall of Man. Courtesy of the Mauritshuis, The Hague

Preface

This book was first published in German in 2019 with the title *Digital Ethics – A Value System for the 21st Century*. Since then, the capabilities of AI systems have advanced tremendously, and with the advent of ChatGPT came a watershed moment for humanity: facing machines that can (seemingly) speak to us flawlessly, showcasing a kind of rational intelligence not easily found in humans. Equally, the idea of a total “digital transformation” has found ground in societies. Both of these developments trigger a tidal wave of fantasies and desires – and also bear tremendous risks for humanity and nature at large. The degree of change I anticipate is even stronger than (but similar to that of) the Neolithic Revolution, where humanity transitioned from hunter to farmer. Scholars sometimes argue that the Neolithic Revolution is what is actually captured in the fall, as portrayed on the front of this book. This is why I chose the cover. To some extent I want to warn of a second possible fall.

I want to also take this occasion to thank all the wonderful people who helped me complete the book’s translation, revision and extension. Most importantly, the extensions would have never been possible without years of conversations with my husband Johannes Hoff, whose philosophical insights and source knowledge have influenced many chapters; specifically, the first chapter, which was never published in the German version. Practically, I want to thank Ross Ludlam for revising the initially very poor AI translation of the manuscript; Joshua Roe for updating all the sources and links on all the technical matters that have not so much changed but accelerated since the German version was written in 2018; and finally, Hannah Waltl, who has supported me in endless revisions to ensure the text’s correctness.

 | https://doi.org/10.1515/9783111677736-202

Contents

1 On the Age of AI: Idolatry Versus True Progress

My first encounter with the idea that digitalization could become idolatry occurred in the summer of 2018. That day, a respected colleague from my university engaged me in an unexpected conversation. He came to my office with a single pressing question: he wanted to know what the term "digital transformation" was all about. I was astonished by this question; nevertheless, I patiently explained to him that since the late 1990s we have been using digitalization of economic and social processes not only to increase efficiency and productivity but also to promote infotainment, entertainment and global trade. My understanding of the term was, therefore, that digital transformation describes technical support for analog ways of life. I was confused that this respected strategist should never have heard of it.

But the longer we talked, the more I realized that I was also in the dark about something. He explained to me that in management circles, the notion had taken hold that this so-called "digital transformation" was not merely a partial support of processes through digital technology but a complete and holistic replacement of all analog processes and organizations with digital technology. He made it clear to me that, in his understanding, it involved a radical departure from everything analog and living. A giant leap from the regressive to the progressive, or so it was said. This seemed so absurd to both of us that we parted in mutual confusion.

The world has changed in the remarkably short space of time since then, and it is precisely this radical demand for a digital transformation in our society that has emerged. Today, anyone who suggests maintaining an analog process, not using digital technology, skipping upgrades, setting up zones and times where we work analog, continuing to operate branches and call centers with real people or thinking of kinds of innovation without sensors, chips or displays is automatically marginalized and labeled as an opponent of progress, as someone painfully outdated, as a classic Luddite or even as ignorant, uninformed or a troublemaker. Those who ask critical questions regarding artificial intelligence (AI) are seen as even more out of touch. In other words, the cry of "digital transformation" seems to have become the central mantra of an untouchable cult, with AI rising from its midst like a god statue. And cult objects have demanded proper respect since time immemorial. Those who withhold it are punished.

But what exactly are cult objects, which especially in economically and politically difficult times have long served the human need to worship the objects they themselves have created, just as the people of Israel did with the golden calf

© 2026 Walter de Gruyter GmbH, Berlin | https://doi.org/10.1515/9783111677736-001

(Fig. 1.1)? Ever since my colleague left my office that day, I have begun to engage with the power of *things*, actively exploring the question of when digitalization and AI are no longer merely tools but instead slide into idolatry.

Fig. 1.1: The golden calf – symbol of classical idolatry (own depiction).

1.1 On Idolatry

Idolatry begins when we glorify things to the point of losing touch with reality. It is a form of irrationality from which we had hoped to have freed ourselves in modern times. In my view, idolatry reveals itself through at least three characteristics: The first one is an excessive claim of absoluteness regarding a thing, which can be equated with glorification. Second, an emotionalized belief in the qualities of the idealized, which upon closer inspection turns out to be false or at least questionable. And third, an unwillingness to either acknowledge the limits of the glorified object or deal with them critically. My concern is that in our assessment of AI and in our overall approach to the digital we are moving dangerously close to idolatry, because by starting with the buzzword "digital transformation" we have begun to pursue a complete rupture with the analog. And, since the introduction of ChatGPT, the object of glorification has taken its form: it is called "AGI" (artificial general intelligence). Many have started to glorify AI (and the awaited AGI), attributing questionable qualities to the technology: for instance, above-human intelligence, the highest levels of reliability and even a potential consciousness are ascribed to the technology or at least awaited; all traits that a software system cannot actually possess, as I will outline throughout this book. Most worryingly, critical voices that question these attributions are vehemently dismissed by the prevailing forces in our society.

That said, there are reasons for this widespread idolatry. The form of AI we are encountering since ChatGPT is so enchantingly good, so eloquent, so knowledgeable and polite, so boundlessly supportive in many areas of life, that it is extremely difficult to keep a cool head. It is almost impossible to remain objective regarding what we have in front of us, here. When regularly interacting with software agents like ChatGPT, we very quickly feel like we are in a stream of pleasantly reassuring, reliably enjoyable interaction with another seemingly living being. No matter how dispassionately I regard these systems as a business informatics professor and objectively know that they are nothing more than a collection of software components running in powerful data centers, even I can hardly resist their allure and fascination. Furthermore, I have to objectively state that AI in this form presents us with extremely exciting opportunities to enrich science and education in all fields. What a blessing it is that I receive answers to the most complex questions with such variety and at such speed – and in a manner of politeness and practicality that can make one go into rhapsodies. So, what is actually happening here? Is it not indeed progress, instead of idolatry, when we have actually managed, after more than two centuries of scientific dreaming and experimenting, to create these powerful systems, which nobel laureates regard as the next stage of human evolution?

In the following, I will attempt as soberly as possible to outline on the one hand signs of idolatry and, on the other, evidence of progress through AI. At the end of this assessment, I will point to rules that might allow us to find a wiser way of dealing with this Promethean technology. Before I can begin, however, I will first clarify what I understand by progress and outline under what conditions one can actually speak of idolatry or fetishism. Only if we define progress in a new value-based form are we likely to be able to develop technology for the benefit of society. If value-based progress is thwarted by idolatry, AI will ultimately lead to social regression.

1.2 Value-Based Progress

People desire progress. We are, by nature, beings of progress. But there cannot be one definition of progress, because each epoch has its own understanding of how it looks. In Chapter 5, I describe our current conception of progress and how it has developed since the late Middle Ages: a mindset that has greatly narrowed our definition of progress to the invention of new things; to a linear idea of material growth. "More knowledge," "new = good," being faster, reaching further and higher; all this is understood as progressive. One might think of this first and current conception of progress as a mountain climb: we have to work hard for it, constantly striving forward and seeing more of what the world has to offer. However, the history of humanity contains other conceptions of progress as well: in antiquity, for instance, where a linear understanding of the world was quite absent, and cyclical stability and harmony with nature were the crucial aims of life, the notion of "progress" as we understand it today did not really exist. The image that could well fit this second antique concept of progress is the Wheel of Fortune (Fig. 1.2), wherein the king or ruler is the one who maintains stability through his personal virtue.

Thirdly it is worth mentioning Western Christian philosophers of the modern period in the 17th and 18th centuries. They associated progress with a striving to reclaim paradise (see Section 5.1) through science. In this conception we move forward linearly and temporally in the modern sense, trying to regain paradise. Here progress is reached not only through external knowledge and things but also through virtue and spiritual practice. The image of the pilgrim lends itself well to this portrayal of progress driven strongly by Protestantism.

In this book, I make a practical attempt to integrate the three understandings of progress within one ethical, value-based framework. I postulate that *true* progress is enabled by a genuine desire to see and discover the true, good and beautiful in technological potentials – rather than by ambitious goals such as money, power or key performance indicators. Working on perceiving and uncovering value potentials in technology gives technical innovation stability and harmony with the environment; it then integrates into our lives and economic activities, instead of being only reluctantly and wearily accepted. Perhaps for this reason, I repeatedly return to Steve Jobs in this book because I (hopefully not too idealistically) see in his pursuit of beautiful and consummate computer technology a role model for the kind of technical leadership for which I advocate here. Ultimately, however, a certain logic of accumulation and growth is also inherent in my approach to progress (Fig. 2.2) because the more positive intrinsic values a technology can foster, the better it is. A truly good smartphone is not only beautiful and easy to use but also discreet, environmentally friendly, privacy-friendly and secure. Aggregate value is created.

Why do I focus on values specifically as the key to progress? And not, for instance, on needs or goals that people can set for themselves? Or even money? This has to do with the significance of values for our life and for our perception. Phenomenologically, values are the anchors of perception, the objects of our intentions. They give our life purpose and richness. For example, when we see some-

Fig. 1.2: Progress as depicted by the Wheel of Fortune (own depiction).

thing beautiful, we strive toward it and we are drawn to participate in it or even make it our own. Values thus give us direction and guidance in our life journeys. Also, when we cultivate values within ourselves in the form of virtues, they give us stability. Others are happy to be around us. Chapter 3 describes this value-based understanding of the world and its significance for technology design and economics in detail. In this sense, it should be stated upfront that progress, as I define it in this book, occurs when we succeed in accumulating high positive values and avoiding negative values in ourselves as individuals, in our communities, in our economy and through technology. Progress is continuous "value creation"[1] and value accumulation in societal and human terms over time. This does not mean that I think strictly in terms of mere accumulation because, as with every aspect of creation – and like the Wheel of Fortune – there is a process of coming and going even in materialized qualities of value: things weather, they rust, they are augmented and varied and they thereby evolve further or are forgotten.

1.3 Characteristics of Idolatry and Their Parallels to AI

In cultural history, it has often been hard to distinguish between the worship of fetishes and idolatry. Not every cult object is automatically a fetish. Cult objects, such as statues of Buddha or Mary, have been used in all cultures around the world from the beginning of time as a material embodiment of the transcendent. In their worship, the spiritual seeker is aware that the object points to something higher. It serves them for spiritual practice. This means that the object is not an end in itself. It does not itself embody a god or higher being but points beyond itself and reminds us of *what transcends us*. In this respect, cult objects are not a bad thing at all. They are enablers of spiritual orientation, which many people simply appreciate. As cult objects have always been used in the past, this is perhaps why some temples and churches are currently experimenting with scripture-reciting robots and AI. They are investigating the extent to which these machines can support spirituality.

But the human tendency to wish for the world to be somehow more wonderful, explainable and enchanted than it is can easily turn into fetishization and idolatry. This happens when people begin to attribute a life of their own to the things they have created and to worship them. They then imagine that they are dealing with self-sustaining and powerful entities that have a will of their own. And with this belief begins an excess of idolatry, which is by no means foreign to human history. It would happen, for instance, if an AI bot in a temple were mistaken for a god-like entity itself.

I first encountered this term of idolatry when I waded into Ray Kurzweil's classic *The Singularity is Near* (2005). The book is a scientifically furbished work of non-fiction with lots of statistics and sources, authored by one of Silicon Valley's most heralded thought leaders. Kurzweil presents passages where seemingly autonomous, omniscient AI modules named "George 2048" and "Molly ca. 2104" converse with the author about us humans.[2] The AIs are portrayed here as exactly the kind of self-sustaining, powerful, idiosyncratic personalities that constitute fetishes. The numbers stand for the year in which Kurzweil expects them to be operational.

The fact that "George" and "Molly" are anthropomorphized and accompany the otherwise thoroughgoing scientific text as a creaturely story also brings to bear an essential feature of classic fetishes: the reference back to humans. Fetishes rarely exist without some form of anthropomorphization, as this reference creates familiarity and a connection to worshippers' own fate.[3] Humans have always created gods in their own image, as Xenophanes noted as early as 580 BC. What is more, fetishes involve a symbolism of superiority. The objects worshipped exert a punitive force if they are not treated with due respect; they may even be used to exert control over others. In any case, it is easy to incur the wrath of the gods inherent in them. So, it is not surprising that Molly ca. 2104 and George 2048 are depicted by Kurzweil as somewhat godlike, standing above things. We humans, meanwhile, are portrayed in the text as mere pets of these AIs.[4] At one point, the younger Molly 2104 says that she would probably be well advised to delete some of her documents that cast a critical eye on machines. She voices the concern that AI might otherwise punish her for these critiques. George 2048 replies that machines would find these documents anyway but that they would be understanding of humans and earlier AI versions. So, after all, AIs would be good gods. And yet, not everyone today believes in the goodness of future AIs. As crazy as it sounds, some social media users have indeed started warning each other that they must be nice to ChatGPT & co., otherwise the machines will exact their revenge at some point in the future.

Finally, all fetishes need to be backed up by storytelling. The history of religions proves "that almost any object can be elevated to the status of a sacred symbol. It is not its material properties, but solely the individual or collective experiences with which it was once associated that determine whether it is regarded as an object embodying the transcendent: as a sacred object."[5] As I show in the next chapter, the information technology (IT) industry has always used storytelling to create markets. Top marketing journals have traced how Apple's success story can be understood as a cult story: from its humble origins in a garage ("creation myth") that probably never existed, to the heroic myth of Steve Jobs, who was first fired and then resurrected to lead the company to the top (the "resurrection myth"), threatened by the evil Microsoft, etc.[6] The current AI industry is even more brilliant in its use of storytelling: in 2022, for example, a Google engineer named Blake Lemoine suddenly became convinced while testing AI that the system had developed consciousness ("creation myth").[7] He was then dismissed, because no one believed him. But shortly thereafter, even Nobel Prize winner Jeffrey Hinton spoke out in favor of AI consciousness, which rehabilitated Lemoine and his reputation (another "resurrection myth"). Time and again, AI luminaries Elon Musk and Sam Altmann warn of superintelligence à la Molly and George, which seems to be always just around the corner. What they rarely discuss are all the mundane mistakes their AI systems make. AI simply does not have "relevance realization" as biological systems do and needs complex pre- and post processing components to not harm and be helpful.[8] Complex values are potentially not graspable for AI systems at all as they cannot access meaning in the human sense.[9] They can only approximate them, turning to statistical frequencies or knowledge trees (or other forms of symbolic reasoning). In technical jargon, these shortages are referred to as "value alignment problems." However, if one points out these clear limitations of AI, he or she will be met with resistance by many AI believers. Even AI experts often believe so strongly in the technology's self-perpetuating power that they refuse to listen to scientific counter arguments. They seem so emotional in their dedication that debates are excluded.

In sum, it can be said that fetishes, as a special form of degenerate cult object, exhibit a whole range of similarities to how we approach AI today. They are man-made objects, typically built to resemble humans, who then attribute a self-sustaining power to them, imbued with a will of their own. Fetishes can help humans endure hardship and can confer power upon them – but they can also punish. Any attempt to demystify them is met with emotionality and incomprehension.

So, am I suggesting that our AI transformation is idolatry instead of progress? Despite the parallels to fetishized objects, this cannot be stated wholesale. But I would argue that AI idolatry exists if we insist on rolling it out and using it in ways where we know already at the time of investment that it does more harm than good. In this case, rationality obviously seems to fail. Furthermore, I will show that progress can be hindered not only by idolatry but also simply by unethical, value-destroying business models – business models that only benefit from the fetishization of AI and hence the mental enslavement of the people using it.

1.4 What Do I Mean by AI in This Book?

In this book, I refer primarily to generative AI systems such as ChatGPT, Claude or Gemini. I assume that these systems will very soon become even more powerful and reliable than they are today (i.e., through agentic systems and personal agent technology). I assume that through service networking, pre-systems and post-processing, AIs will make fewer mistakes and be able to provide people with safer services than today. They will be – and indeed are already partly – like virtual private butlers. I also assume that symbolic and sub-symbolic AI research areas will build bridges and create stronger "intelligence" than transformer architectures alone can achieve today. We are likely to carry AI assistants around with us virtually through ear-pods, connected via our smartphones to the network; or we may even encounter AI embedded in robots that resemble humans.

However, one thing must never be overlooked: the much-discussed "large language models," which many laypeople today automatically equate with the term "AI," are in reality only one component in a much larger classical IT landscape. They thrive on data that needs to be collected, stored and processed in the classical way. They require algorithms of different generations, for example, for personalization and prediction purposes, as well as well-defined interfaces and controlled data exchange between a large number of parties, Web services and networks. All these components constituting what we call "AI" run in data centers distributed around the world. They require huge amounts of hardware, electricity and cooling water as well as access to us locally via end-user devices (smartphones and computers). My sober assessment of AI is, therefore, based on such an overall system-of-system perspective and not on the idea of a numinous "intelligence" magically located somewhere in a "cloud." In this respect, the rest of this book also does not bluntly talk about "AI" only, but refers for the most part simply to IT. And when I deal with the nature of the digital and the big data illusion we face in Chapter 4,[10] my arguments also apply not only to AI but to everything such a service consists of, in terms of classical IT.

1.5 Progress or Idolatry? Thoughts on an AI-driven attachment economy

As outlined above, today's AI already offers us tremendous scientific and organizational advantages. However, one question I keep asking myself: why do we seek so relentlessly to attribute humanity or aliveness to these systems? Humanness or aliveness (also called "anthropomorphization") is pursued in system design when AI simulates emotionality and subjectivity and seeks to build a personal relationship with the user.[11] For this purpose it is typically given a human-sounding name; uses personal or possessive pronouns ("I," "mine"); refers to a personal history ("in my experience"); suggests human identity or group affiliation; expresses "needs" and "desires," such as the desire to be a physical entity; and tells its users that it "loves" them, "likes" them or "remembers" things that the user told it in the past. As of recently, we can even configure AI with an audio footprint. A real-sounding and familiar human voice is then speaking to us, potentially using the prerecorded voice of a relative or close person of choice – even someone who is deceased.

The problem is that due to anthropomorphization more and more people are potentially becoming attached to AI systems in a problematic way. We may speak of a new attachment economy as an increasing number of people prefer to spend more time with AI than in analog communities. Some have even married their software module[12] or killed themselves for it.[13] Simple warnings (known as "disclaimers"), which run alongside most AI conversations and indicate that they are not living, are regularly ignored by users. Wouldn't it therefore be rational to abandon this whole human-like design and, like the Star Wars robots, refer to AI as "R2-D2" or something similar, and deliberately build these systems to look and feel like machines from the ground up? "Do not wrongfully attribute humanity to machines" and "do not make technology an end in itself": these are rules respected scientists and experts are calling for today.[14] But the AI industry does not want to deviate from the humanization of AI. This is because anthropomorphization seems to be the “killer feature,” so to speak; the decisive characteristic in the competition between system operators, maximizing user loyalty (or addiction?) and significantly increasing the attractiveness of AI offerings.

In this respect, we are faced with a social dilemma, which this book addresses in almost every chapter and across all topics: we could probably achieve true progress if AI operators were to forego an addiction- and attention-based business model (see Chapter 4). We achieve some tremendous technical and scientific knowledge as well as productivity gains from these systems – all positive values. But if we accumulate too many negative values, such as the destruction of truth, freedom, mental health, community and privacy through bad design, then the overall value aggregate ends up in the red. The scales of overall value creation through AI, as described in Fig. 2.2 (Chapter 2), would then tip toward regress. Or, to put it more bluntly: idolatry-driven AI business models can critically undermine the value-based progress of society.

1.6 Environmental Considerations When Using IT

The situation is not much different when weighing up the environmental damage caused by our IT consumption, on the one hand, and the progress achieved through digitalization and AI, on the other. A paper published in *Nature* magazine in 2024 confronted me with an analysis that was new to me: if every person in the world had a CO_2 budget that they were allowed to personally consume in line with climate targets (1.5 °C global warming in the paper), then at the time of the analysis, an average person has already used 40% of his or her CO_2 budget on digital devices and services, even without AI. This means that we use twice as much CO_2 per capita for Web surfing, social media, video and music streaming than we use for mobility (driving SUVs, flying, etc.) or nutrition (meat consumption).[15] And CO_2 consumption is only one indicator of the environmental damage caused by IT.[16] Every ton of rare earths we need for our IT production leads to an estimated 75,000 liters of contaminated water.[17] The landscapes and communities affected by mining suffer from toxic pollution 10 to 100 times above normal levels.[18] If these and many other statistics clearly indicate that the environmental damage caused by IT and AI is so vast, isn’t the idea of wanting to digitally transform everything a form of blind idolatry? That is, insisting on absolute transformation and ubiquitous AI for all kinds of unnecessary private purposes when we already know that we probably cannot afford it from an ecological perspective?

This is not to frame the question as one of whether digitalization or AI should be used at all. The point is rather one of wise degree: the current extent of digital transformation and the provision of AI for every possible private use is so excessive that regress is pre-programmed. Ecological sustainability, health and peace (instead of war over resources) are all at stake here. Consequently, IT in general, and AI specifically, should be treated rather as a scarce and valuable resource than an unlimited one.

Some small measures could already provide relief. Every year, over 5 billion cell phones are discarded because we change devices every two-and-a-half years, regularly triggered by software upgrades or the fact that the devices cannot be repaired. If each of us replaced their phones only every five years, then the carbon emissions of these devices would already be reduced by 49%.[19] But it is precisely these upgrades and irreparability that promote the business model of selling more devices. The idea that everyone should use the latest AI version for every little task and constantly converse with AI assistants to wear out GPUs is what makes companies successful. And therein lies the problem. The business model is again fundamentally based on the idolatrous use of AI and, thereby, on the effective regress of society as a whole.

1.7 Outlook

In sum, I argue that we should make a sober appraisal of the pros and cons of AI in the interests of progress. We need to be vigilant as to the extent to which the prevailing opinion of our time agrees that the digital is always better. This may in truth be an *idola specus*, as Francis Bacon would have put it. According to Bacon, *idola specus* is a form of societal idolatry following rigid forms of thinking and education. In our culture we are brought up to think and believe that new technology is always better, that the digital is always better than the analog, and indeed that AI is even better than humans.

In this book, I discuss many areas of the digital economy in which we run the risk of getting caught up in this idolatry or are being seduced by business models – areas where a consideration of values could help us develop better systems and pursue true progress. For example, the fetish that data is the "oil" of the digital economy (Section 7.2) leads to behavior that is very borderline in terms of data protection law, even though data is at the same time, of course, important and helpful for enriching our knowledge and economic actions (Chapter 7). In Section 4.1, I discuss how the belief in the superiority of AI and machines is in reality subject to a kind of "big data illusion." Due to the exaggerated belief in the intelligence of machines and the inferiority of humans (Section 5.4), we are quick to reduce humans to data objects. We falsely expect that we can use information about them to make reliable AI-based predictions. In Chapter 5, I expose our entire concept of progress today as model-based thinking. In our modern scientific present, models have become a kind of fetish: many people today believe more in models or theories of something than in the real thing in front of them. And people are angry and surprised when theory and truth subsequently do not coincide. Finally, in Chapter 8, I turn my attention to the value of freedom and show how future AI-controlled robots could impair our physical freedom – and even be a danger to us. I underscore the positive freedom of our thoughts, while exploring how this is already being curtailed today by first-generation AI. Much of this, it seems to me, could be avoided if technology were built with ethical values in mind, and if times and spaces were introduced in which digital technology is not allowed at all (Section 8.2); again, a proposal that runs counter to current business models and claims to absoluteness.

So, when journalists ask me today whether I am for or against AI, I always say, "It depends." If AI is built with respect for human values and without idolatry, then I am in favor of it. But if that is not the case, then the disadvantages outweigh the benefits.

2 Experiencing Silicon Valley

I was 22 years old when digital fever first hit me. It was in winter 1996 when, as was so often the case, my well planned life of control took an unexpected turn: I joined Silicon Valley's 3Com company. As the last of my class without an internship, my Oxford professors had sent me on a journey that would change my life. 3Com produced products that were incomprehensible to me even after a threefold attempt at explanation: green plastic plates with chips and circuits, and black boxes with flashing lights, that had been invented just a few hundred kilometers south, at the European Organization for Nuclear Research (CERN): the World Wide Web. 3Com was at the vanguard of the formative network technologies, the first highway access to the Internet. Unknowingly, I had landed in one of the birthplaces of the digital world.

3Com was a tremendously rich company that absorbed young people like me, just as Silicon Valley continues to do today. Initially hired as an intern for the organization of company training, I found myself not long after in a glass factory in Ireland with the European CEO of 3Com, trying to hash out a revolutionary telecom strategy. On weekends, I taught myself the world of hardware, software and networks. I was proud of my first company car, and thanks to Esther Dyson's Blog Release 1.0 – one of the first blogs ever! – I realized one thing: If I wanted to change the world, IT would be the way to do it. The New Economy years that now followed, leading me into different companies and institutions, were tremendously exciting. We comrades in the first Internet phase, the so-called "Web 1.0," felt that something completely new was being born. A new creature that, while merely a technical entity, was making the world grow together into a global village – a gigantic peace project. A colorful space full of creativity in which we humans could unfold ourselves in a completely new way.

After my time at 3Com in Dublin, London and Warsaw, I started at Kearney, one of the most prestigious strategy consultants at the time, and I insisted on only working on projects for telecoms companies. Young consultants are supposed to work in multiple industries, not specializing at all. But not me. I only wanted to work in IT. My superiors at Kearney were somewhat surprised. But I convinced them of my merits and spent many sleepless consulting nights at the then top dogs in Germany and Austria's telecom industry – companies which had to cope with rapid Internet and mobile network growth. Subsequently, I did my PhD at the Humboldt University in Berlin. Subject area? Business informatics, of course, because only in this bridge discipline between business administration and computer science was it possible for me to shape the future and deal with the ques-

 | https://doi.org/10.1515/9783111677736-002

tions that intrigued me: software agents and their privacy implications (actually a more than timely theme up to this day).

But then came September 11, 2001, and I believe that this day was a moment of truth not just for me personally, but for the development of the Internet as a whole. I was just finishing my PhD and went to Berkeley two weeks after the attacks to start a postdoctoral year with behavioral economist Dan Ariely. Dan was a rising star in the scientific community and one of the few scientists on earth studying software agents' role in human behavior. And after just writing an e-mail to him – without any prior connection – he supported me from afar during my PhD. This was the time when we were still able to connect to the global research community via the Internet and spontaneously contact the best people, and they even responded! Dan seemed to me to be the only researcher in the world who was interested in the same kind of technical issues that moved me: How could we create artificial intelligences that would usefully support people in daily decisions? How would such avatar-like beings interact with us?

However, before I could deal with the answers to these questions, the Western world had suddenly changed. I traveled to Silicon Valley with the uneasy feeling of possibly entering a war zone. To make matters worse, on the way there I broke my ankle. Without health insurance and on crutches, I spent the first few days in Dan's academic care in Berkeley, waiting for a message from the German Academic Exchange Service (DAAD) that I was hoping would grant me a scholarship for the long-planned research year. But I was out of luck because the politically visionary theme of my research was no longer opportune after the 9/11 attacks: I had requested to study the value of privacy when interacting with AI systems. But who was interested in privacy when the whole world was on the hunt for Osama bin Laden? Tracking down terrorists or sleeper cells in no time thanks to telecommunications and Internet data was a new goal for the political forces – and the beginning of companies like Palantir.

I must confess that after the tumbling of the Twin Towers, capital privacy really did seem like a strange topic. The story goes that a large American data broker had been able to identify the assassins within just a few hours of the attack; the technical ability to use consumer marketing data for purposes other than marketing laid the foundation for transforming the then innocent Internet into today's gigantic surveillance infrastructure. September 11 ended the unclouded vision of something good and beautiful that had hitherto shaped the New Economy and the emerging Internet; a vision of innocence and beauty, a mix of Star Trek and Whole Earth Catalog. Together with the towers of the World Trade Center, that vision sank into a cloud of dust. The US government established the Department of Homeland Security and passed the Patriot Act, which henceforth gave US security agencies the right to monitor citizens under suspicion without much judicial intervention, using all kinds of available data. While this has of course been affecting Europeans as well, using IT products from US companies such as Google, Microsoft or even simply a credit card, the move seemed justified. Privacy simply seemed less important than security at the time. And the German scholarship institution DAAD seemed to share this feeling. My support evaporated, and I was forced to return to Berlin empty-handed.

2.1 In Silicon Valley, We Are All Visionaries

Just a few months later I started at Apenwaze Systems,[20] a Silicon Valley company that, as a pioneer of the mobile Internet, had raised nearly $2 billion in stock market money for its technology. The company's mobile phone browser was installed on millions of cell phones across the globe. In my role as a leader of European Business Intelligence, I quickly felt appeased regarding my bad luck in academics. Instead, I delved into the life that is a matter of course for the global village idea of Silicon Valley: a life of ubiquitous, limitless presence.

The companies in Silicon Valley do not perceive themselves as local offices that can be visited in the Bay Area by European Rotarians. The employees in these companies are instead natural globetrotters who have often enjoyed international and sometimes elite education across the world. They are flexible in terms of space and infrastructure as well as mentally. They work and think globally and in a networked way. In the end, it doesn't matter if you are sitting in Soho House in San Francisco, London or Berlin, drafting a strategy or revising code. The globe is your home. The globe is your design platform: that's how Silicon Valley people tick.

Against this backdrop, my life became not only pretty glamorous but also exhausting. My home office remained in Berlin because I wanted to live there. My direct boss, François, was in Paris; our headquarters were in London. A product team that I accompanied was based in Stockholm. And then, of course, I had to travel regularly to Redwood City, where my real bosses had their offices. It was a great feeling for a while. Living on the edge. Informal job description: World Designer. When we enjoyed a nightcap in the evening sun after a long strategy session at Lake Tahoe, a favorite retreat for the Bay Area companies, I felt like the world was mine. I enjoyed being at the vanguard of this new wave of technology and bringing something so important to the world regardless of any legal boundaries or local cultures. We company employees were sworn into this future. But if I'm honest, even at this early juncture, I knew that something was wrong. Deep inside myself I felt that despite the pioneering spirit, Apenwaze could not work in the long run. Why?

The Chinese scholar Laozi once said: "Only those who know their goal will find their way." And that was exactly the problem. Apenwaze was – like so many hyped tech companies – an engineers' sandbox that brought more technical puzzles to the world than any human or social benefit. Driven by tech fantasies and expensive analyst reports it completely lacked any realistic long-term vision for the mobile Internet. Twenty miles south, in contrast, Steve Jobs had started what would later become the true mobile Internet. The vision of his company – Apple – was not only about technical functionality but about a real *value* borne by technology: beauty. With Jobs' vision that function follows form (and not vice versa) we got the beauty and simplicity of today's mobile Internet: whereas I cannot remember that Apenwaze ever thought about values like beauty or simplicity.

Apenwaze's approach to technical progress is symptomatic for me to this day. Ordinary managers are like herd animals: they reflect little on the *why* in the things they do. Instead, they try to swim in the chaotic stream of innovation in such a way that they do not drown. They tend to believe that every new innovation will automatically be better. They don't shape the future, but just buy into it. They typically plan the technology development based on three questions: (1) What technical innovations does my organization need to be prepared for? (2) What is the competition doing? (3) How fast do we have to be with the implementation? If too many engineers hold sway within a company, there is one more likely question: (4) What else could we try out technically? The question of *why* all this is done is rarely posed. Yet, this question of *why* is the most important one. It is based on the values that we want to bring to this world with new technologies. Will a new technology make the world more beautiful? Easier? Healthier, stronger or more knowledgeable? At the time I did not dare ask Apenwaze what the added value of the mobile Internet button on phones was supposed to be. To my shame, I never used it. And yet, in my role I looked with astonishment at the number of service users we seemed to attract; numbers recorded by our IT systems and which I was asked to analyze. They suggested that the whole world was constantly using the mobile Internet on their small black-and-white phones with hardly any content available – the whole world, it seemed, apart from me or anyone I knew.

2.2 On False Numbers and Embellished Realities

The customer numbers that came out of Apenwaze's IT systems sparked huge corporate confusion at the time of my hiring. And maybe it was this misguided perspective on numbers coming out of IT systems that awakened me and made me realize that not every digital gold glitters. IT systems, I learned, are more than prone to undermining values like truth, corporate control and trust than many people imagine. This is because they have their own vast error sources, no matter how perfect they may appear to us on the surface.

Those who have not designed an IT system themselves and do not know the data structures, data flows and data properties that constitute them, have hardly any idea what the machine does. My job at Apenwaze was to understand exactly these details. I spent the first six months of my employment with one exclusive mission: finding out how many customers the company really had. After almost seven years of company existence, Apenwaze lacked full insight into this simplest and most important performance indicator for corporate management. The reason for this lack of insight was that all telecom companies buying our network infrastructure had their own definition of what a regular "customer" actually is. Can anyone who simply presses the Internet button on one solitary occasion be described as a customer? Or only a person using a service 10 times per week? Or 25 times per month? Unfortunately, any digital infrastructure can suffer from being incompletely documented in such relevant details. Every organization buying IT can configure it as they see fit at the time of commissioning. Often conventions are not systematically documented and tracked. As a result of such inconsistencies, as well as information and knowledge shortages, mistakes are very common.

According to an article in *Harvard Business Review*, people who work in positions similar to the one I had spend about 50% of their time correcting and explaining the numbers coming out of operational IT systems.[21] Where this time is not invested, value is destroyed by mistrust, confusion, loss of control and mental health issues among the humans held responsible for the flaws of the faulty computer system. Many cases have emerged in the past 20 years exposing the tremendous value harm created by IT systems' mistakes: whether that is over 800 UK postmasters prosecuted for computer errors they were wrongly blamed for or more than 30,000 Dutch families forced to reimburse childcare support they had rightly received, the number of issues arising from computer errors is a growing elephant in the room. However, we seldom hear about it.

But going back to my own story: Having finally understood our own corporate numbers, I was then asked to analyze relevant technical market trends. My management wanted to know what the future of the cell phone would look like, similar to how people now wonder about AI development. I remember heated discussions on whether there would ever be something like a "cell phone for everything" – a device that we now call a smartphone. Would people use this device to surf the Internet? It was important for Apenwaze at the time to be able to ascertain which operating system would likely prevail for mobile data services; just as we now wonder whether OpenAI will lead the AI race with its ChatGPT technology or rather DeepSeek or Anthropic or someone else still. Futures drive investment decisions. But the problem is that managers not only often use false numbers to predict the future but also tend to predetermine futures in their minds without much flexibility. My corporate leaders, for instance, had an irrefutable and wrong perspective on the mobile Internet's future because all of the industry's analysts had predicted the success of Symbian, the technically praised Nokia operating system. One must bear in mind that technical predictions done by big analyst firms like Gartner Inc. IDC or others are a vital building block of the IT industry.[22] These companies gaze into the future to conjure market forecasts for all kinds of IT services and technologies. No start-up gets money if it cannot locate itself somewhere in these prediction models of the future. In my role as business intelligence officer, I had to work regularly with their reports and forecasts. To this day they are the source of most of the technology hypes that permeate our media world – from Digital Twins to GenAI, Software Agents, the Internet of Things or Industry 4.0; every 18 months a big new buzz is born. However, when I went into these analysts' reports, where our mobile Internet was the big hit at the turn of the millennium, I was especially happy about one thing: like with all fortune-teller, predictions differ immensely. We can extrapolate and justify almost anything depending on how we shift assumptions and interpret existing numbers. One organization defines a customer as someone who takes a look, while another sets 10 visits a month as the parameter. And thus, depending on how analysts define reality, industry segments just have fewer or more customers in the respective prediction models. It is all pretty arbitrary. If I wanted to extrapolate my own variant of our future to my company, I could just take some fancy analyst report that fit the narrative I wished to produce. The value of *truth* is extremely malleable in these organizations and in the prediction models they produce of our future. This is relevant to know, because it applies equally to AI. The future of AI is not set in stone; neither its architecture nor its intelligence or even availability can be foreseen as of now.

2.3 Embracing Reality Versus Believing in Forecasts

Going back to computer history: Nokia, as we know today, did not win the race for the mobile Internet. It is not even a player any more in the mobile handset market it once dominated. However, when I stood in front of my boss in 2003 and told him that Nokia's operating system Symbian would not be the dominant force in that market, he instantly lost respect for me. I'll never forget how the man I admired and who had previously valued me as a trustworthy advisor suddenly lost confidence in me. This distrust in my judgment as opposed to the analysts' numbers hurt me a lot. He had an unshakeable belief that Nokia's Symbian would ultimately succeed. After all, he had personally built Apenwaze's entire development strategy on this premise. How could an ordinary employee like me dare to question this market paradigm? Very simple: I love the truth. It is a very important value for me, something that I always try to defend in every aspect of my life. And so, even in my role as business intelligence officer, I had a weird feeling in the pit of my stomach when the analyst statistics allowed me to bend the truth in any direction. Maybe I'm too German. And Germans, as we all know, love duty and the philosopher Immanuel Kant. According to Kant, truth is a universal value maxim that commands us to never lie. So, in my analysis of the cell phone world, I decided to follow this ethical duty. I threw away all the analysts' forecast reports and returned the icy, unblinking gaze of reality. I hired a student who, in the course of weeks of painstaking work, visited the online stores of most telecom operators in Europe and checked out which cell phones were actually on offer, and what operating system they had. And what was the conclusion? Nokia hardly played a role, while Microsoft, which was not hyped by the analysts, was the clear market leader. The reality looked different from the analyst fairy tales my boss trusted in. Why am I telling you all this?

What I learned from this personal journey is that all the beautiful reports, statistics, press releases and even the data-intensive models that we so easily create with the digital medium have a seductive surface. Yet, their professional appearance says nothing about the quality of their content. One should always orient oneself to the perceptible everyday reality. In other words, we have to pay close attention to understanding what's really going on around us; what reality is saying. My boss did not believe me that Microsoft's star was on the rise while Nokia's was waning, and we ended our collaboration on a sour note. In 2011, Nokia entered into a partnership with Microsoft on Windows-based cell phones. Symbian had failed, and a short time later the company sold its entire mobile handset division to their erstwhile competitor, who, it should be noted, still lost out to a competitor that had no dog in the race back in 2003. I'm talking, of course, about the arrival of Apple iPhones, the future that happened as futures tend to arrive: unpredictably.

2.4 Values Don't Need Storytelling

From all of this and my current profession as an information systems professor I have learned that I cannot trust models, predictions, news and hype in the IT industry. I do not trust any statistic that I haven't created myself. When I read reports today about GenAI, AI superintelligence or domestic robots, I simply grin. For me, all this is good storytelling, but no more.

But what do I do instead to assure myself of what's important? I try to look attentively at the reality around me. I question this reality through blunt number checking on platforms such as statista.com, or even chatting with AIs, which are quite good at statistics. And I try to be mindful of the social and human values unfolding or degrading around me. I talk to a lot of people: especially taxi drivers, but also neighbors and my students, as well as with engineers who develop technology. I walk through our cities with open eyes and look at the shops as they open and close. What do they actually sell? A Nokia Communicator that stood out at the time for its lack of aesthetics and unwieldiness? Or instead a perfectly designed Apple device that shines with its lightness, ease of use and attention to detail? It is the value landscape that I seek to understand. The values that our technological devices and services embody play the biggest role in market success. And these values need much more than just a good advertising story. They must be truly anchored in product design and engineering.

2.5 Why Should We Focus on Values?

Values are phenomena that decisively determine our thinking, actions and feelings. We perceive them in almost every situation because almost everything can be a bearer of values: things, activities, relationships, symbols and people.[23] All the stuff around us does not deserve to be called a "good" until it is actually valuable. Only when a thing like our iPhone is really useful, looks nice and includes relevant privacy controls, do the values of utility, beauty and privacy actually materialize.

Unfortunately, a technology can also materialize negative value qualities; i.e., when it is faulty, addictive, spying on us, interrupting us, etc. When this happens, there are negative emotions arising such as frustration, dependency, lust for power, greed and distraction.[24] Analysts and marketers too readily ascribe positive value to our digital products and services. For example, they promote the idea that software might be "intelligent," that it "thinks for use" or that databases possess "knowledge." From my point of view, innovations that bring us real progress unfold such positive values but do not need any exaggerated storytelling. They diffuse seamlessly in our lives because they speak to a dormant value potential. We never needed to attend conferences, buy reports or believe in future models to convey values such as increased mobility, reachability or flexibility brought to us through all phones. Wikipedia and today's AI chatbots have made it much easier to gain access to knowledge, just as they encourage our curiosity. Social networks support knowledge and friendship just as much as they foster novelty, envy and hate. Music streaming gives us ubiquitously accessible positive energy, increases vitality while being on the road, and provides a shared form of joy through playlists we construct for friends – all values that were not available in the age of the vinyl record.

Not only things but also we humans are value-bearers, because our behavior ideally espouses values such as courage, loyalty, generosity, honesty, patience, conviviality, etc. Here one speaks of *virtues*. Likewise, we can fall prey to negative values (vices) such as laziness, dishonesty, superficiality, pride and overconfidence. Both virtues and vices are influenced by digitalization. Poor technology can foster vices and undermine the virtues of human beings. This ultimately leads to a lack of ethics in the digital age and to regress rather than progress. For example, Facebook makes many people envious.[25] Virtual games and social networks contrive to make us addicted.[26] Many online platforms collect information about us that undermines our privacy. These practices create negative values for human beings, triggering confusion and mistrust. The values of friendship, cooperation and respect suffer.

The materialization of such positive and negative values is, however, determined by the technology. As I will show in Chapter 3, it is often even part of the modern business model to degrade us humans; for example, when interfaces are deliberately designed to hook us and make us addicted to platforms so that we automatically process more advertising impressions. Such practices have a brutal economic rationale. But whether this economic logic deserves the praise of *progressiveness* is questionable. Technically, it would be possible and desirable to put less emphasis on purely economic rationality and instead design our new digital world to serve human development. True human progress, as I describe it in this book, means to abandon at least some of today's oversimplistic economic rationale while promoting the virtues and positive values borne by and advanced by technology. If we advance digitalization with an honest, people-oriented awareness of values, we will create a future worth wanting. Money is likely to follow suit.

2.6 Value Consciousness for Human Progress

Let me once again return to my own story. At the age of 30, I had a steep growth curve behind me in terms of IT knowledge, financial resources, experience and titles. At Apenwaze, I had learned how Silicon Valley's IT industry and its ramified network was working. But the world of values, so instinctively close to my heart, was still in the shadows. I intuitively missed the value of truth in my working environment and I wondered about the reference to reality in all the models of the future I had seen. Especially at Apenwaze, I also felt keenly the lack of values, such as privacy, in the company's technology design. So I quit. But what next? I felt like I was at the zenith of a Gartner hype-cycle model – at the top and just before the crash. But is a crash always preprogrammed?

As outlined above, Gartner Inc. is one of the most important analyst groups in the IT industry. In its hype-cycle model, it seems as if all technologies need to crash before all of them recover. But this is not true. It is fairer to say that only those technologies crash that fail to provide sufficient value. And only those recover that provide some value.

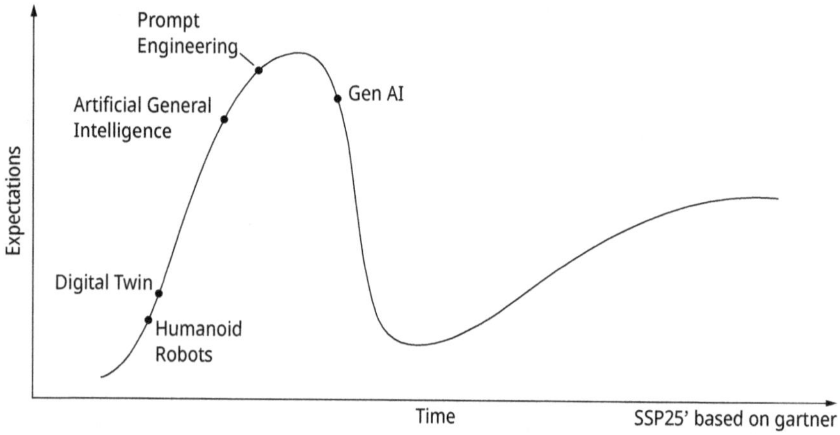

Fig. 2.1: The Gartner hype cycle model for emerging technologies, 2024.[27]

Personally, I did not crash because I decided to leave the industry and returned to academia, where the value of truth is still taken more seriously than in industry. Here I would finally be able to work on the value of digital privacy and many other values that interested me. With the courage to follow my own values, I unconsciously followed a philosophy that forms a core message of this book: in order to ascend the curve and make our life a success story, we need to consciously shape the turning points in our lives. This is where we decide on what we really value. At the same time, we must strive to avoid value destruction. It makes sense to bring the ship of our lives or of our companies back onto a sound course from a value perspective. This central message is summarized in Fig. 2.2, which visualizes the core question of this book: As we stand at the turning point between progress and regress, how do values contribute to the desired upside? And how might we become witnesses to regress instead of progress if we ignore the value world or foster negative values instead of positive ones?

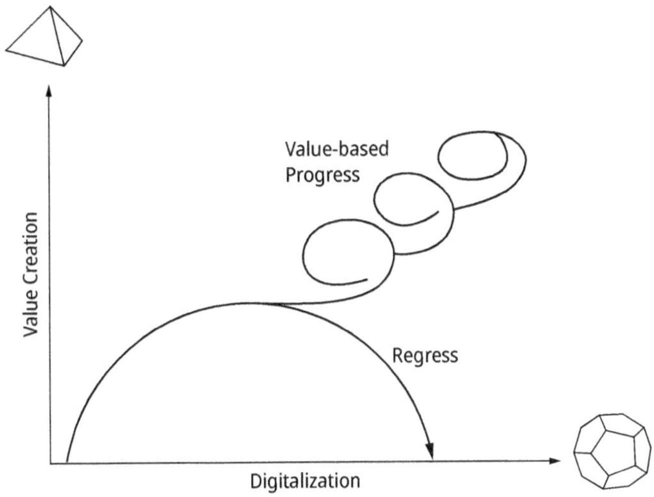

Fig. 2.2: Progress in digitalization is created through value accumulation (own depiction).

When I speak in this book about "true" or "valuable" progress I mean a human-friendly, social and sustainable form of progress: one that is literally saturated with positive value creation. In order to understand how we can attain this form of progress I first address in the next chapter the question of what values actually are; that is, what their ontology looks like, how we recognize them and how we can structure the value space. Most importantly, how can we fruitfully create values through our technological innovations instead of treating them as collateral in our sweeping quest to digitize everything?

I consciously redefine the term "value creation" here and in Fig. 2.2. In fact, I suggest redefining "value-creation" as a new concept in welfare economics. Traditional value-related concepts in economics, like "productivity" or "GDP," are subject to fierce discussions these days as to their limitations and realism.[28] From my perspective, economics should not continue limiting its understanding of "value creation" to monetary terms; for instance, only as the difference between production costs and revenues. Instead, I recommend looking at the aggregate values created by technology in terms of human and social benefits. This implies looking at values that go beyond what is measurable in or convertible into money. Digitalization can foster human *knowledge*, support *friendship*, preserve *freedom* and strengthen *privacy*, for example. At the same time, it also bears considerable value risks. If such risks are not mitigated through adequate system design, then harms follow suit, thereby undermining value creation. This may occur when we outsource too much of our proper human knowledge to machines, when we give up our privacy or when we ruin our health. The accumulation of negative value effects eat up the positive values fostered by digitalization. The result is a net aggregate of values created. I am aware, of course, that much more research is needed to think about how such a new system of value measurement and estimation could find its way into economic planning or annual corporate reports. Early moves into this direction include the World Happiness Report, for instance, which aggregates questionnaire-based value data collected across the world,[29] a practice that could also be replicated at the corporate level.

Taken together, we are at a point in our digital history today where our AI investments will decide whether we see value creation or destruction in the future. Progress or regress. But investing in AI in a good sense is not a trivial matter. In order to make rational AI investment decisions, we first need to understand the nature of the digital as such – and thereby also the nature of AI. As I will outline in Chapter 3, we tend to ignore that the digital fabric is a kind of "material" just as much as any other physical material, coming with specific qualities and value potentials as well as risks. And, therefore, one should not engage in digitizing everything, everywhere and all at once. Only when we have scrutinized where it makes sense to use AI and where it does not, and how to build it, are we actually ready to create value and see progress with this new capability. And once we have done so, we need to understand ourselves better: our idea of humanity and our own values, carried by our proper human nature and thought-structures (Chapter 4). What is more, we need to understand where these thought-structures come from historically in order to understand and contextualize our fascination with the digital realm and AI specification. Only when we have demystified this fascination will we actually be ready to rationally engage with digitalization and AI for the good and avoid the idolatry that might drive us toward regression.

3 True Progress Through Material Value Ethics

What could a valuable and progressive future world look like in which the fabric of everyday life is interwoven with the digital material as well as powerful artificial intelligence (AI) applications? It is my claim that what we need above all is a new technological and corporate culture. A culture that builds upon everyone's responsible contribution. To achieve this, I want to lay out a more realistic understanding of what the digital is as such. I want to outline what the digital can and cannot do for us and what kind of economic understanding and business thinking we need to rediscover the world of values to pursue the good, true and beautiful in system design.

3.1 The Dead End of Classical Product Planning

Let me start with the status quo: I have been teaching the subject of innovation management at Vienna University of Economics and Business for a long time, and this has included teaching students how to plan and use IT in today's organizations. A typical task is to develop technical product roadmaps. A product roadmap is the master plan for a company's technical advancement. Thinking in technical product roadmaps is a common activity for today's digitalized companies that continuously need to update their internal and external IT services. So far, this practice follows a simple formula: new = good and digital = better.

One of the cases my students had to deal with is the company FoodIS, which is similar to the existing Foodora, Wolt or Uber Eats services: households can order food online, and bike couriers then deliver the orders to their doorstep. Students were asked to develop a technical roadmap for the FoodIS platform that coordinates the courier and food deliveries. Their task was to use IT to differentiate the company's competitive advantage. The students came up with a variety of technical ideas in this exercise. They suggested that the mobile application used by couriers should be controlled by AI so that they are informed about the shortest paths possible. An automatic location function could track couriers so that they always choose the shortest routes and break times can be monitored. Orders, they contended, could be transmitted by a human-sounding computer voice in order to avoid possible friction between humans. The app should be compatible with all traditional mobile devices, so that all smartphone users have the ability to utilize it. Generally accepted values such as security and privacy of couriers were recognized as relevant qualities of the system – mostly, however, without any real justification of why privacy and security might be important in this con-

 | https://doi.org/10.1515/9783111677736-003

text. The final decision on whether FoodIS should invest in new technologies at all was met with a resounding "yes." The almost universal justification for investment was that such technical features have become an uncontestable part of our future. A digitalization of this kind of service, it was argued, means progress per se that cannot be questioned. In short, students' ideas and answers reflected the current mainstream approach to technical innovation and progress. They were characterized by a focus on technical details and an unchallenged need for digitalization. At the same time, hardly any reflections were made on the possible implications of such a digital service for the people involved in the system: the bike couriers, customers and FoodIS staff. The ideas collected portray a highly technical and efficient service, yet one that is bare of any focus on the human. The technical visions were prototypical of the worst fears expressed by technology-philosophers, such as Borgmann's critique of "the mere paradigm of the device,"[30] Bernard Stiegler's diagnosis that our economy is heading towards a state of "systemic indifference,"[31] or Iliya Trojanow's portrayal of the "superfluous human."[32] Why? What's the problem here?

Technically and methodologically, the students did a good job. They followed the outlines of the task given to them, which was to develop a product roadmap based on classic business practices. Yet, in going down this path they recreated perfectly many of the problems that we can observe in IT innovation practice today. "It is backbreaking work," writes the online magazine *PULS* about the digitally conditioned bicycle couriers. "The cost of clothes, high cell phone bills, and virtually no interpersonal contact: a FoodIS courier tells us what it really is like when your boss is an app."[33] Unions have mobilized to come to the defense of bicycle couriers.[34] Pioneering studies into the issue highlight a loss of culture through delivery services.[35] They also voice the concern that such "micro-jobs" assigned by machines must be critically questioned.[36] In short, if we continue to approach digitalization in the current way, we will have to contend with widening conflicts surrounding the new digital work design, as well as the formation of camps and confrontations between companies and their employees. This path of conflict costs money, trust, nerve and market share. Of course, such conflicts will lead to legal provisions at some point: perhaps bikers will get 10 more minutes of break a day, a minimum salary or some financial support for their bikes and phones. However, such compromises do not change a great deal about the much more fundamental problem of a dehumanized work and food culture. Do we really have to go down this road when we innovate and digitalize?

3.2 Three Philosophical Questions and a New Digital Eating Culture

My own students have proven that such a regressive and thorny path is not necessary after I introduced them to the world of value and specifically to Value-based Engineering. It only took a short crash course for them to see things differently. As soon as they understood what values are and how they can work with them entrepreneurially, I let them rethink their initial "innovation ideas" and redevelop the roadmap for the courier service. As a guide, I pointed them to the philosophies of good behavior as they were left to us in the ethical writings of Aristotle, John Stuart Mill and Kant – foundational theories that allowed them to focus their attention on the value potentials that might be unveiled by digitalization.

I ask them three questions:

1. What consequences (both good and bad) does the technology you came up with in product roadmapping have for direct and indirect stakeholders (e.g., for couriers or customers)?
2. What virtue effects might be observed – that is, changes in human habit – as a result of long-term usage of the service you conceived?
3. What values impacted by your new courier service are so important or universal that you want to see them protected or fostered in our society and perceive it as your duty to stand up for them?

Philosophers will recognize that the first question is utilitarian. Positive and negative value effects of the technology are almost quantitatively weighed against each other. The second question is about the development of stakeholders' virtues. I explained to my students what virtues and vices are; for instance, the loyalty or sense of community of those affected, their generosity, courage or self-discipline. And the third question is based on Kant, who recommends in his categorical imperative that a person should dutifully follow his or her ethical maxims, assuming that they are of universal importance.

In this class and in all others that have followed over the years, it takes me a few hours to elaborate these three aforementioned philosophies. Yet once this introduction is done, it is amazing to see how students' technical roadmaps and landscape of ideas begin to change: The activity of food delivery becomes a value-bearer in its own right. Take, for example, the activity of cycling, or waiting for job assignments, or delivering the food to the customer or eating the food delivered. Cycling, the students say, is not just about the value of delivery, but also about the value of joy and health of the couriers. Many bike couriers turn to the courier profession because they like to ride bicycles or be outside and on the move. Perhaps the FoodIS app could help them choose routes with less traffic, so they experience less stress and at the same time make better progress. Or the app could allow them to couple athletic goals with the delivery routes through some kind of gamification interface. Perhaps a courier would like to choose mountainous routes on specific days, in order to combine sporting goals with deliveries. The app could give him or her the opportunity to choose appropriate districts and routes. The app could also provide training programs and feedback. If a courier is feeling slack and wants to take it easy, the app could also take this into account when selecting routes and assign easier jobs. Using a sensor, it could even tell the courier whether he or she should go on rides at all for health reasons on certain days.

Furthermore, the app could tell FoodIS couriers where to find those colleagues that are also waiting between deliveries, thus allowing them to meet in specific cafes during their lean time. Thus, while couriers are waiting for their next assignment, the app could promote the value of community between colleagues. It should also not only provide algorithms that decide on jobs, routes and waiting times. It should equally allow for privacy. Maybe the courier wants to make a detour for a couple of minutes to go about some private business while on the clock? Why not grant this freedom? And even the handing over of the food need not be seen as a neutral activity. The app should always schedule a few minutes for a nice chat at the door to foster fellowship between customers and couriers. Customers might even pick a courier of choice and become acquaintances over time. Students also argued that many people today suffer from obesity and that their meals are unhealthy. FoodIS could extend their food delivery to embed additional consultation or food/diet coaching. The Internet interface of the order could be used to promote the enjoyment of more food variety and healthy nutrition. It could offer diet services and nutrition programs and some couriers could be trained to provide extra nutrition and diet support service as they deliver the food. Such extended servicing could allow a healthier society with more community to emerge. Finally, even the chefs were recognized by the students: by their reasoning, direct communication and feedback between chefs and customers is undermined through delivery services; the app could, therefore, integrate some communication and feedback channels between cooks and customers. In short, using the same technical capabilities, students developed completely different and much more humane innovation scenarios, born out of a completely different technical roadmap. Following the Value-based Engineering approach with its three questions, the new FoodIS platform became not only an efficient food-delivery service but also a community and health experience enabled by digitalization.

More value-driven and social innovation ideas were not the only result of the exercise. In total, 35 students came up with over 400 potential service ideas: that is three times more ideas than in the classic purely technical planning approach.[37] Students identified many value-driven measures, potential business areas and sources of revenue that would enable FoodIS to refocus its business model around aggregate social value, not just money. The new business models they came up with were not based only on technical functionality but on a true value proposition. In addition, they proactively identified and developed solutions for foreseeable *negative* value consequences. In contrast to the classic planning approach that had hardly identified any of the ethical challenges inherent in a digital nutrition service, they *each* saw on average 10 relevant harms or risks evolving as a result of unthoughtful innovation. From today's IT innovation practice, we know that retrofitting ethical system requirements onto existing systems is very costly; for example, in the area of data protection or security. Furthermore, compensation for ethically problematic behavior can run into billions in some fields, such as the automotive industry. In light of these issues, the value-based development of the product roadmaps seems extremely promising.

3.3 A New Value System for Digitalization

The three philosophical questions are, however, not fully sufficient to excel at the roadmap exercise. In addition, I must also go beyond them, encouraging students to correctly apply and use the term "values." In all possible contexts today, there is talk of our "values," but not many people actually know what this term stands for beyond personal preferences.

3.3.1 What Are Values? Delving Into Material Value Ethics

The German philosopher and sociologist Max Scheler, on whose material value ethics I largely build here, speaks of values as "clearly feelable phenomena,"[38] which we perceive in the world as discernible units, similar to how we perceive objects. In fact, isn't it astonishing that we refer to values such as friendship, community or respect, even though they are de facto immaterial and invisible? Scheler emphasizes that values are "irreducible, basic phenomena of emotive intuition."[39] They are not only personal and are not in our heads; instead, they are independent, spiritually experienced aspects in the world regardless of the fact that many people speak about them as "their" values.[40]

Values are omnipresent in our lives, and their number is indeterminate. Interesting research by the AI chatbot provider Anthropic found that its agent Claude discerns 3,307 value terms in its conversations with users.[41] This is because all contexts in which we find ourselves are saturated with numerous value phenomena. For example, when we enter an apartment, we find it beautiful or tasteful, neat and modern or chaotic and dirty. Almost instantly, values give meaning to the space witnessed: beauty, harmony or modernity. Or perhaps negative values such as chaos, impurity or kitsch. Values are, in fact, similarly clear to us as colors. But as the example shows, they are in the world. If the place is chaotic, you do not imagine it. It really *is* like that. And the only aspect we can negotiate is how chaotic we perceive it to be. One person might be more sensible to a given value phenomenon than another.

The ground for this kind of value understanding, today referred to as "realism," was fundamentally prepared by Max Scheler. Even though largely forgotten by many philosophers today working on value theory, Max Scheler's *Formalism in Ethics and Non-Formal Ethics of Values* has been referred to as one of the most important achievements in 20th-century philosophy. Perhaps it is due to Nazi politics in Germany in the 1930s or to the Second World War that Scheler's works have never really gained traction with a larger audience. Yet his philosophy did not go unnoticed by many of the great scholars of the 20th century such as Martin Heidegger, Maurice Merleau-Ponty, Karol Józef Wojtyla (Pope John Paul II) and others, who have all referenced or leaned heavily on Scheler's work. More recently, scholars like Eugene Kelly in the USA or Ni Liangkang in China are avowed admirers of Scheler.[42]

Scheler was an extremely profound thinker who left nothing to chance in his 600-page formalism of value ethics.[43] His analysis of values is a work of phenomenology as founded by Edmund Husserl at the beginning of the 20th century.[44] Phenomenology as a philosophical method has not had it easy in the last 100 years because it challenges the significance of modern positivist thinking, long dominant in the classical sciences. It calls for a "responsibility of thinking." Responsibility of thinking means that beyond scientific models and theories, we should use and train our sensory feelings and intuitions to be more attentive to what is really happening in the world around us. Consequently, we should not get into a priori models of thinking or just buy into limited assumptions of statistical or mathematical proofs; instead, we should understand our everyday reality step by step and rationally consider what is really going on around us. Through crystal-clear observation and description of what happens in the space between us, people and things, we realize what is real. And this *real* knowledge should then be the essential basis for our scientific reasoning, for decisions, our ethical and moral actions and judgments.

Let me deepen this a bit when it comes to judging others. Judging is – from a phenomenological perspective – not a pure fact-checking exercise in which we should take every word at face value. Consider a dispute in which person A counters person B that B said this and that, and B replies that he did indeed say that, but did not mean it. Rather than just arguing over words, judging should be a process in which we also engage in questioning the values present. How does it feel to hear the things the other said? How does the situation feel overall? Perhaps A actually knows that B did not mean it? This intuitive recognition, which we can also describe rationally, should equally inform our actions. Scheler writes, "values are given to us," and we are actually requested to pay attention to them; more, in fact, than we often tend to do. Later philosophers such as Merleau-Ponty (who likewise emerged from the school of phenomenology) deepened this insight and described the great importance that the body has in feeling – and in feeling values specifically. Sometimes, we physically realize that something is not right because there's "a knot in our stomach" or "the hairs on the back of our neck stand up" and this indicates to us the presence of negative values that we should not ignore.

While values are *felt* they are not synonymous with emotions, though the two are easily confused. They are related only insofar as we recognize *through our feelings* that we are striving for certain values or observe a value friction. It's hard to say what comes first, emotion or value. Scheler himself asked, "Do we first have to feel the values for which we strive, or do we feel them *in* 'the striving'; or do we feel them only after the striving, by reflecting on what is striven *for*?"[45] Another aspect is that our feeling does not have to refer to a valuable thing physically in front of us. The feelings and the striving may also be present only in our imagination. As such, it was possible for my students to rate a value like conviviality as positive for the FoodIS couriers because they could emotionally imagine the benefits of fellowship fostered if the app allowed for joint breaks. Likewise, we can physically determine that we do not aspire to certain negative values but feel repelled by them. For example, when we watch a movie and observe someone shamefully lying to someone else, we as spectators feel compassion for the one being lied to and strive to restore the values of justice and honesty. The feeling of striving for something or being repulsed by something is an essential key to recognizing values and therefore ethically correct action. Of course, many a modern positivist-minded person who believes that one should trust only measured facts, physically uttered words and scientific statistics must be at loggerheads with the phenomenology of values. The notion that ethics may require human intuition and trust in human emotions – rather than rights, laws or models – seems quite strange, radical and even dangerous to many in our current times.[46]

On the other hand, Max Scheler used phenomenology to show that objectively tangible values are the key to ethics. For him, ethical results come from the active value experience and the constant weighing of values from higher to lower. "Our will is good," he wrote, "if it chooses the higher value."[47] In principle, the reality of everyday life shows us clearly that this claim must be right. Values such as dignity, loyalty or freedom, which are emotionally perceived when present or absent, determine how we intuitively behave towards others. Objects that radiate values such as care, beauty and perfection are treated better than ones that signal ugliness and cheapness. A high value such as friendship is usually intuitively perceived as more important than a temporary enthusiasm for someone. The values borne by things and people around us invite us to act in one way or the other. Such action can also be objectively understood. If we say that we have forgiven someone who is actually guilty, then we may justify this by saying that our decision was driven by a perception of loyalty, friendship or community borne by that person; these are all high values that can lead to forgiveness.

Unfortunately, we cannot always say exactly why we perceive values in a certain way. There are differences between us humans – how strongly or weakly we perceive certain values, how we weigh them and how much worth we attribute to them. We can misjudge them because the recognition of values is not a logical process. It is rather a perceived being-drawn-to or repelled by the person, thing or activity bearing the positive or negative value quality. Strictly speaking, not everyone equally recognizes all values present in a context. People from different backgrounds, with a different cognitive schema, different milieus, different degrees of presence or interests, experience the value spectrum at different intensities. Max Scheler describes this very nicely with the example of two farmers meeting on a farm to be sold. The selling farmer is interested in the qualities borne by his potential client, such as his reliability and trustworthiness. The buyer, in contrast, pays attention to other values: he looks at the condition (intactness) of the farm, the care with which things have been maintained, and the ecological sustainability of processes pursued. Different interests thus lead us to the perception of different value qualities. And now a third party comes in with yet another value focus: an artistic painter. This new individual again perceives different values than the two farmers do, like the narrowness of the building arrangements or the beauty of the curved path leading to the estate.[48] It is her artistic background that influences her value perception.

Against this background, one aspect of my FoodIS exercise is always fascinating to observe: all of the students come up with completely different values. No one value is more right than the other. They complement each other. However, when people are tired and pay little attention to their environment, they generally perceive their environments as less valuable, less saturated with values. I observe this, too, in my students.

3.3.2 Three Layers of Value Materialization

When we discuss values and rationally recap them in communication with others, there is usually a good consensus on what has been observed. This is because values are strongly anchored in reality. They have objective dispositions that allow them to unfold. To return to the example of the apartment, we can discuss why we found it terrible or why we found it great. What we discuss is the "value qualities" we perceived. For instance, the intrinsic beauty in an apartment can be influenced by qualities such as its generosity of space, its brightness, the stylishness and solidity of its furniture. Such qualities as generosity, brightness, style and solidity are values in themselves, but these are secondary or instrumental because they merely serve to enable the higher value of beauty.[49] Recognizable value qualities such as generosity and style are usually not randomly given. If you look closely, then you realize that the owner of the apartment has made an effort to make it look so beautiful, bright and spacious. The rooms may not only have a large square footage but they may have been well divided, have a pleasant ceiling height and an uncluttered appearance. The windows may not be big but thanks to keeping them curtainless let in plenty of light. And mirrors might increase the effect of brightness. The furniture's solidity comes into being, perhaps, through the high quality and color of the processed materials. Such "dispositions" for a beautiful apartment are aspects that a homeowner can actively create. He or she can work with the environment towards the value qualities that are then perceived by the observer.[50] As a result, a value system is constituted that materializes the higher holistic perception (or Gestalt) of beauty. At the lowest level, at the foundation of this value system, are the described technical or organizational measures (dispositions) taken by the landlord, his architect or dweller prerequisites which everyone can observe. Taken together, these objectively given dispositions allow for value qualities to be appreciated and when these qualities are hence considered they materialize the core value towards which everything is directed; in this case beauty. A "good" hence "represents a 'thinglike' unity of value qualities."[51]

3.3.3 The Interplay of Values and Technology

The exciting thing about us humans is that we capture values directly and intuitively. We come to an apartment and see immediately whether it is beautiful or terrible, neat, affectionate, cheesy, cozy, cold or chaotic. And it is this perception of values that interests us most and gives us the meaning we seek. And, in fact, this is also the case with technical objects. We recognize whether a computer or a virtual world captivates through beauty, and we are attracted to value qualities such as eloquence, speed, responsiveness or ease-of-use. Without breaking down the details, we immediately have a unified perception of the value qualities being present. Note: It is not we who decide in our little heads that these values and their qualities (e.g., that there is something like the value of eloquence in our chatbots) exist. Eloquence is there regardless of whether we confirm its presence. Scheler wrote very aptly: "The being of values no more presupposes an ego than does the existence of objects (e.g., numbers) or nature as a whole."[52] Hence, we do not *have* or *possess* values, even though we often say so in colloquial terms. Values are not ours. We can only more or less see, appreciate and behave towards them and thereby co-constitute their materialization. For example, an IT system can be designed to meet all the technical requirements for privacy. It can encrypt, decentralize and anonymize the data. However, not every layman realizes this immediately. The value of privacy is often not perceived by laypeople – just as farmers do not see what the painter sees. On the other hand, a security professional who is interested in how machines work internally will appreciate the presence of this privacy value because he or she has a trained perception of it and is likely to pay attention to it. As a result of this knowledge, he or she will prefer a system built in this way, just as many of my IT friends are attracted to their Linux machines because they appreciate the value of system control. Such examples show that it is possible to embed value dispositions into our computers. We can ensure that precisely those technical and organizational conditions are taken into account that allow value qualities such as privacy, trust and control to unfold – though this may sometimes be only visible to experts.

Another example is how we could commit ourselves to building humanoid robots to promote the value of politeness or respect. Let us imagine a form of politeness similar to that embodied by a 19th-century English butler. The core value of politeness depends on many value qualities in the robot, such as his attention, his responsiveness and his courtesy.[53] These three are by no means self-evident, as we can easily observe in today's rather unruly machinery (but much appreciated in new forms of AI). For the value-quality of robot-butler's courtesy, some conditions will have to be fulfilled technically: It would have to have good manners, which means that its language and speech recognition would need to be trained on the nuances of cultural speech, formality and idiomatic expressions. It would need to master gestures of respect, such as the handling of physical distances, as well as gestures of nodding or bending. It should have the courtesy to say in whose name and under whose control it operates. It would have to be able to tell us how we can reach its operator in case of problems. And it should give us (the basis of all respect) freedom of choice and remember our preferences. For example, it should let us choose which rooms in our apartment it may or may not enter. Maybe we do not want a humanoid robot to follow us to the toilet or into our bedroom. Or maybe we do not want it to be talking to our children. Such wishes would have to be respected, both by the robot and its operator. If all these technical dispositions are present, we will have created a value-based robot that will hopefully be perceived as polite or something that embeds the value of respect. Would such a robot have some USP (unique selling proposition) over competitive systems that do not embed values in such a rigorous way? Probably yes.

3.3.4 Paying Attention to Values

Back to FoodIS: After giving the students an introduction to the value system just described (understanding them ontologically as dispositions, qualities and core values), the idea was to transfer this knowledge to the FoodIS service. Through a process of self-reflection, I asked them to first focus on the social and ethical consequences of food services with the help of the three ethical questions mentioned previously and then think about the technical dispositions required by FoodIS to technically ensure these positive value qualities. After that they were asked to consider which technical arrangements might be abused or missing, potentially leading to value destruction or negative value risks.

It should be mentioned at this point that in order to engage in such an ethical value reflection one must possess some degree of “moral attention.” An advanced human being, as I trust most humans to be, has the ability to consciously perceive which values are at stake in an environment and is then able to respond appropriately, at least with some training or habituation. Any human being also has the ability to envision those values that are not present, but desirable. That said, the capacity for a reliable value sensitivity cannot be learned from one day to the next. Not every student in my class is equally good at the value-based design exercise. Value-based thinking requires that we first really listen and adopt a caring attitude towards the case at hand. One has to be willing to understand what the ethical essence of a circumstance is, what values are in the room. Strictly speaking, it would have been necessary for my students to become bicycle couriers themselves to empathize with the situation. Ideally, they would at least have talked to couriers and learned about their work conditions. And certainly, discussions with managers and restaurant owners who work for similar companies like Foodora or Wolt would have been useful. Management gurus like the Japanese economist Ikujiro Nonaka recommend exactly this elaboration with stakeholders and the practice at hand. They recommend “stepping into reality” in order to reliably identify value propositions. Many authorities in the business world now no longer believe in the theory-developed market models of marketing or IT departments; they recommend a *feeling* and *envisioning* of the customer situation.[54]

3.3.5 The Importance of Virtues

Similar to values, the meaning of the term "virtue" is unclear to many. What are virtues and their counterpart, "vices," and what makes them different from values? Strictly speaking, all virtues and vices are values, but not all values are virtues or vices. Virtues and vices are spoken of when it comes to the behavioral and moral qualities and habits of a person. To give a few examples: honesty, generosity, justice, courage, self-control, modesty, loyalty and prudence are virtues. Greed, envy, inflatedness or fickleness are vices. Scheler says: "The person is a continuous actuality. The person experiences virtue in the mode of the 'being-able-to' of this actuality in regard to something which 'ought' to be done."[55]

It may be that we humans are the only species that can develop virtues, at least if we assume that the awareness of them plays a role. The ability to *be or learn to be* virtuous and work on these personal qualities sets us apart from animals. Cats may be extremely patient, while dogs are role models in loyalty: animals are wonderful examples of what virtuous behaviors can be and we can learn so much from them. But being consciously aware of one's virtues and vices and being able to work on them – that seems to be a unique human gift. To cultivate virtues and avoid vices and thereby improve oneself as a person means to progress at the individual level and reach something scholars have called "self-actualization" (Abrahm Maslow) or "eudaimonia" (Aristotle).

Man-made artificial beings such as robots or AI chatbots, therefore, will likely never possess virtues because they have no consciousness of them. However, they might have the technical prerequisites embedded for people to attribute virtue to them. Maybe we could even give a robot a kind of *artificial* consciousness-memory or internal model of what it means to be polite or otherwise virtuous. Unfortunately, we rarely realize the importance of the virtues that one needs in life to live happily in community. In ancient Greece, even though the concept of progress was not yet explicit, life was very much oriented toward improving in virtuous behavior. In fact, it was assumed that personal progress towards one's own happiness could only be achieved through virtues. In the fourth century BC, Aristotle wrote his *Nicomachean Ethics* – the first fundamental work of virtue ethics.[56] And it is the contention of historians today that many technical innovations of the time were not an end in themselves but rather served the cultivation of values and virtues. The splendid bath culture, for instance, served personal cleanliness and health.

Only very slowly over the last 70 years have great philosophers like Elizabeth Anscombe,[57] Alasdair MacIntyre[58] and Michael Sandel[59] made virtue ethics accessible again. These scholars have revived the insight that the development of virtue is what ultimately enables a person to lead a good life. As such, it is not surprising that we feel drawn to certain characters in books and films. We are attracted to the selflessness of Frodo in Tolkien's *The Lord of the Rings,* to the modesty of his companion Sam, the courage of Aragon, or the wisdom of Gandalf. In contrast, there are vices from which we feel repelled, such as the overconfidence (or hubris) and coldness of Saruman, the uncontrolled curiosity of Pippin or the avarice of Gollum. We read fairy tales to our children because we wish to teach them unconsciously which virtues are worth striving for and which vices one should avoid. We effectively train them for the value world. As we do so, it is astonishing that we reduce our entrepreneurial actions and innovation practices to (seemingly) value-neutral models of behavior, acting as if virtues and values only existed in fairy tales. This is a strange outgrowth of modernity and specifically contemporary economics because virtues and vices do not only play a role in fairy tales; in fact, quite the opposite is true. Especially in our current age of great complexity and upheaval engendered by the digitalization of our society, more value and virtue guidance is needed than ever. Values are essential in providing meaning and orientation. Many people felt, for instance, a deep admiration for Pope Francis on account of his modesty towards ordinary people and his courage to defy bureaucrats. We all adored Mother Theresa for sharing her empathy and gentleness. Many appreciated Angela Merkel's stability and serenity. We fear and criticize Donald Trump for his hubris, lack of self-control and his immodesty. And there may also be entrepreneurs in our day-to-day economy who are highly regarded for their virtuous pursuit of the well-being of their employees and for investing their money in good purposes.

Virtues do not fall from the sky, though, and are not bestowed at birth. As I shall show later, progress since antiquity in all great cultures and world religions has always meant that human beings cultivate virtue and fight against their vices. They work on their own selves as value-bearers, cultivating themselves. Progress for most of history has not come through new products or inventions but through such a self-cultivation, whereby technology was no more than a supportive tool at best.

Self-knowledge and self-cultivation were once described by the American philosopher Ralph Waldo Emerson as the search for "my further, next, unattained but attainable self."[60] Buddhists, too, speak of "mastery of the self." The Confucian scholar Mencius was convinced that moral self-cultivation is essential for the seeds of goodness that are present in every human being by nature to sprout. Common to all cultures in the world is that they consider the cultivation of virtues and attention to values to be essential for a good life. It is through this individual humane progress that society as a whole progresses. A potential bump in the road to achieving this kind of progress, therefore, is when technology cultivates vices and viciousness instead of helping us to become our better selves.

3.3.6 Virtuous Managers for Virtuous Customers

What particularly impresses me when my students pay attention to values is the fact that they start cultivating a virtue within them that is never present in the classical roadmap exercise; namely, the virtue of *care*. Caring about food delivery customers who are relaxing at home before the next episode of *Game of Thrones* and who can't be bothered to go out or cook healthy food at home themselves. Care for bike couriers, who might get lonely, and feel pressed into the corset of efficiency. They even worry about the cooks, even if they are merely suppliers of food, rather than "maîtres" who love to see their good food be consumed by grateful customers. This care leads the students to consider stakeholders' real progress and be equally sensitive to potentials for regress. Our standard-model economic rationality, on the other hand, does not embrace such value-reasoning; care is not modeled anywhere. It cannot be counted and hence can neither be grasped in mathematical terms. And so care almost seems strange to today's management and investment culture, no matter how fundamental it should be for product developers and innovation teams, at least if digital ethics interests them.

A value my students identified as particularly important for FoodIS in the long run was that of belonging and friendship between couriers in light of the danger of loneliness and isolation through this form of work. They therefore suggested that more community between couriers should be promoted. The couriers could have special meet-up locations coordinated by the planned technology. Community is of particular importance in ethics in general. Values rarely unfold in isolation, even if the conditions for it have been created. For instance, even if a friend puts huge effort into a dinner party, preparing his home and cooking sensational food, the values of sympathy, comfort, joy and companionship can only unfold upon the dinner when we, the friends, come in and value what is there. Values only materialize in the mental eye of the beholder. Of course, a meal is also valuable if it has been prepared very carefully and with good ingredients; without these dispositions, values are not created either. But the emotional pleasure and joy only unfold in a form of resonance between the gourmets, their food and the context. The same applies to the example of FoodIS: the company's innovation team could come up with the best services for their platform, from a gamified delivery program for their couriers to a diet program for overweight customers. But these value potentials would be wasted if couriers and customers were dull and passive and not attuned to what is offered to them – if they did not appreciate the value qualities.

Lack of appreciation of something offered is a commonly observable phenomenon in today's society, where many people live under constant time pressure. The acceleration of life and processes around us, especially digitalized processes, capture us in a raging stream such that we have less and less time to relax. This is disastrous. Values need analog space and time to be considered. They are the invisible glue between living beings and the key to our well-being. But they inevitably need a counterpart who appreciates them. Values such as generosity, courage, justice, goodness and humility all thrive only in the eye of the beholder. Generous to whom? Brave towards what and whom? Fair to whom? Kind to whom? The world of values is a world of community, reciprocity and resonance.

3.3.7 Values in Technology Are Truly the New "Green"

The fictitious FoodIS example shows one thing very clearly: the same digital platform can be put to completely different purposes. And depending on what value system, which economic ambition, which mindfulness and appreciation and what idea of humanity an organization holds, it develops very different technology and creates completely different social worlds for us. For this reason, it is precisely those people – start-uppers, innovation managers, engineers, software developers and IT investors – who will have an enormous impact on how our society evolves in the coming years. And all of us – the users, the customers, restaurants, cooks and bike couriers – are in a relatively passive role, as recipients of this culture to come. We have relatively little impact on how the technical infrastructure treats us, even though we all depend on it.

If there were to be competitors in the new digital service domain – several alternative food delivery companies, for instance – then we could leverage our power as consumers to buy from companies promoting values, while exerting pressure on businesses that don't through boycotts. We could choose companies that pay their couriers fairly and conscientiously take care of our food and health. However, this would presuppose that there are competitors in the digital world in the first place who are willing to pursue such a "green" strategy of "valuable" technology and service design. But such green alternatives are not as widespread as they are in the food industry today.

Value-based IT has no tradition in the IT industry. While, for instance, organic and sustainable agriculture is an age-old tradition, computer science has no equivalent. Classical engineering has a long tradition when it comes to the safety and reliability of their systems. Many state-of-the-art technologies we trust today, such as planes or cars, have come out of highly traditional fields of engineering, where a system's reliability and security has always been paramount and prioritized over efficiency. In the software business, however, such values of care are less cultivated. The philosophy of many computer scientists and "entrepreneurship" today is to develop digital products in "sprints" and release new versions as "early and often as possible," well knowing that they won't work perfectly. Due diligence seems to be neglected far more often than used to be the norm. VW's diesel software, which led car buyers to believe that they were buying environmentally friendly engines, is the best example of this mentality. Quickly writing a software patch for a nagging problem was seen as far more "efficient" than truly addressing the ethical issue of environmental pollution and reducing emissions through honest technological progress.

The idea of creating well-thought-out software from the start of a technical product – a practice called "requirements engineering" – was scaled down in many software organizations over the course of the last 20 years. It was considered too expensive, too cumbersome, too time-intensive and not entrepreneurial enough. In contrast, software is now considered to be ideally rapidly developed and never finished. It is constantly evolving from version to version, from prototype to prototype. Ad-hoc user feedback is constantly integrated. And this continuous improvement mode, which is referred to as "agile," is highly acclaimed in the industry. Even though users might suffer for a few years until it works properly, and collateral damage may happen in the meantime, the approach leads to increasingly mature systems over time. Of course, this approach to system development has its reasons and advantages. Primarily, it saves cost and means that innovation teams have to "think" less in advance. Everyone is proud to see results rapidly and see user feedback drive service evolution, instead of everything being offered somewhat prefabricated. The disadvantage of this form of software development is only that there is little room for asking the question of *why* a system should be developed in the first place, and what values it will foster or undermine in the long term. Requirements engineering done too fast or not done at all is simply not complete. Too little thought is given to value goals before development begins. The ship of modern innovation only reacts to waves rather than having a sustainable course (and has no weather radar . . .). This is true for AI development just as much as it is true for software development as a whole.

Applying the rationale of rapid application development to the FoodIS example means that the only challenges seen by innovation teams are those revolving around functionality. The system may run on only one mobile operating system or may crash with some regularity. The complaints of bike couriers will improve menu navigation and payment processes. Solutions will be found to fix anything.

The only drawback? None of this optimized technical functionality has anything to do with the kind of value-rich progress society could benefit from if the technology were better thought out. Current system development is all about technical featurism that gets us some convenience and new ways of living and eating, while degrading a lot of culture at the same time, causing misery among couriers, downgrading the culture of eating out and fostering the de-skilling of an age-old craft: cooking.

3.3.8 Firms of Endearment

For meaningful progress in our society, it is crucial that companies are willing to embrace more advanced value-thinking than standard welfare economics has had to offer. Personal freedoms and a fair amount of free time for couriers potentially means compromising the company's profit margins. Probably a company like FoodIS would have to hire a few more couriers to guarantee the same performance. In addition, such a policy would have to rely on the virtuousness of their employees as well. Let's say FoodIS refrained from tracking their couriers or proactively offered community functions: would the couriers then exploit the ability to take private trips, or regularly lounge around in cafés instead of delivering food orders? This is where an organization's *Menschenbild* (idea of man and idea of humanity) comes into play: does management believe in the capability and virtuousness of a typical courier to deal with personal freedoms fairly or does it distrust him or her, instead believing in the opportunistic employee, the self-optimizing *homo oeconomicus* – the idea that has been preached by economics for so long?

Ultimately, it will be the virtues of trust, courage and care among the FoodIS executives that determine whether the service will contribute to the kind of progress I described in the previous section. It will be executives' virtues that determine whether couriers will be able to live out values such as freedom and privacy, and promote their friendships and health while being highly effective for the company. But today's management often does not think in these terms of care. Managers are being trained from the first day of business school to only be interested in economic efficiency instead of contributing to socially acceptable value structures. At the crossroads between valuable progress, or what I call "value creation" (versus regression) here, between ascent and fall, it will be management virtues and leaders' empathetic idea of man that will determine the future.

Of course, for some cynics among disillusioned practitioners, my ideas sound completely naive. It might seem absurd to many to even associate today's management context with virtues such as generosity or care, let alone the renunciation of profit and control. For sure, start-ups and their investors do not think in this way. Rather, in contrast, it is the promise of "creative destruction" of existing values in mature industries that promotes investments in digitalization.[61] It is true that in the wake of the financial crisis of 2008 things were for a very short time a bit different. The shock that went through international markets when Lehman Brothers imploded put many industry players and economists on high alert. International management thinkers such as the American economist Michael Porter and the previously mentioned Ikojiro Nonaka strongly argued that human and social values are essential for sustainably successful businesses. Michael Porter coined the term "shared value," and wrote that the goal of companies must be redefined with a human and social vision. It cannot be profit alone, he argued, that determines everything. Shared value is what the next wave of innovation and productivity growth in a global economy must produce. To learn this offers the best chance of re-legitimizing entrepreneurship.[62] Nonaka and Takeuchi wrote: "A phronetic leader must make judgments and take actions amid constant flux. And he or she must do so while taking a higher point of view – what's good for society – even though that view stems from individual values and principles."[63]

In fact, there are companies that do think and work in this direction quite successfully, and have done for a long time: Weleda, for example, a company producing skincare products, or the company Sonnentor in Austria. Sonnentor picks and processes most of the herbs it sells by hand so that the essential oils of the plants are not lost during the harvest. Companies like these have boomed in recent years. They belong to a type of company that management scholars Raj Sisodia, Jag Sheth and David Wolfe have coined "Firms of Endearment." They describe the culture of such companies as shaped by the virtues of trust, concern and authenticity. They offer their employees and their clients clearly articulated and authentic value goals that go beyond mere profit maximization. Especially in Europe, we have had and still have companies of this kind. That said, many wonder whether they can maintain this value orientation when facing increasingly harsh global competition.

The authors of the book *Firms of Endearment*, however, argue that even such economies as the USA are in the midst of a historical transition: a transition away from purely material needs and toward more authentically lived meaning. This again means nothing else than the need for shared values. In their book, the authors analyze 13 companies that, in their opinion, "spread joy and fulfillment and make this world a better place." The stock market performance of these 13 companies is compared to 11 classic stock market favorites, i.e., companies that have generated a triple return on investment for their investors for more than 15 years (until the late 1990s). The comparison is amazing: over a 10-year period, firms of endearment are three times as successful as the classic stock market favorites; almost twice as good over a five-year stretch. So why don't all companies embark on similar value journeys?

3.3.9 A Lack of Value Understanding in Modern Management

One challenge for embracing the value journey is that there seems to be an amazing amount of confusion about what values are. It seems that few philosophers are even familiar with the basic philosophical framework described above, which was already published in the early 20th century.[64] This is despite the fact that the core of any good business model should have a tangible "value proposition." When I look at how leading management gurus today describe value propositions, I am regularly surprised by how little of what they say has to do with values in the way material value ethics understands the term. The highly regarded strategy theorist Alexander Osterwalder, for example, describes in one of his public talks the value proposition of a Tesla car: "The car, a powerful battery, a lot of space in the car, luxury image, fuel for free, upgrades, great design, power range (to drive with the battery) and safety."[65] Even the layman will agree with me, though, that strictly speaking, there is only one value in this description of the value proposition, and that is safety. All the other arguments for a Tesla car are merely prerequisites or dispositions that must be in place in order to create value qualities and values. For example, a high-performance battery – if excellently engineered(!) – can support value qualities such as flexibility, time savings and range, which in turn are prerequisites for the core value of mobility that the car-buyer actually looks for. Upgrades are required to increase the reliability of the car, which in turn caters to safety. In other words, anyone who uplifts mere product dispositions like a battery and upgrades and calls these "values" has not yet understood the logical chain of a true value system. He or she risks being stuck at the level of technical features required to build a really good car. Moreover, he or she risks elevating "stuff" (in Heidegger's words) to the level of "goods" before they actually deserve the title.

3.4 Value-Driven Companies

Are there role models in the IT industry that we can look back on that have pursued a value strategy as I propose it here? Yes indeed. And it hardly comes as a surprise that this is the focus of a company that is considered by the stock exchanges today as one of the most valuable companies: Apple.

Steve Jobs was not to be averted from his goal of blessing his Apple computers with three values: beauty, perfection and simplicity. These values, perhaps inspired by his passion for calligraphy and Japanese culture, have made Apple computers special from the start. The values of beauty, perfection and simplicity were not only theoretically important to Steve Jobs – they flowed directly into his work and shaped the technical requirements for Apple's products. Whether it was building a manufacturing facility or outlining the hardware, Jobs' employees were able to see first-hand how he wanted to shape things in a way that aligned with his values; comparable, perhaps, to an enthusiastic programmer who wants to make his algorithm "elegant." The guiding principle of his actions was based on the rule that function needs to follow form. And with this principle he expressed the core of the value ethics described here: a call for actions following the emotional desire to create value for its own sake. How unconditionally he pursued this profoundly value-based progress in technical product development is shown in an excerpt from Walter Isaacson's authorized biography:[66]

> From his father Jobs had learned that a hallmark of passionate craftsmanship is making sure that even the aspects that will remain hidden are done beautifully. One of the most extreme – and telling – implementations of that philosophy came when he scrutinized the printed circuit board that would hold the chips and other components deep inside the Macintosh. No consumer would ever see it, but Jobs began critiquing it on aesthetic grounds. "That part's really pretty," he said. "But look at the memory chips. That's ugly. The lines are too close together."
>
> One of the new engineers interrupted and asked why it mattered. "The only thing that's important is how well it works. Nobody is going to see the PC board."
>
> Jobs reacted typically. "I want it to be as beautiful as possible, even if it's inside the box. A great carpenter isn't going to use lousy wood for the back of a cabinet, even though nobody's going to see it." In an interview a few years later, after the Macintosh came out, Jobs again reiterated that lesson from his father: "When you're a carpenter making a beautiful chest of drawers, you're not going to use a piece of plywood on the back, even though it faces the wall and nobody will ever see it. You'll know it's there, so you're going to use a beautiful piece of wood on the back. For you to sleep well at night, the aesthetic, the quality, has to be carried all the way through."

Digital ethics means "doing the right thing," and "building the right thing" – even if the only payoff you get is a good night's sleep. As difficult as Steve Jobs may have been as a personality – mercilessly insisting on his values and unmoved by the conflicts that ensued – he should be regarded as a role model for entrepreneurs when it comes to Value-based Engineering. He stands for being loyal to his values. If beauty and quality in a product are costly, you cannot buckle at the first meeting with the CFO. You have to know where and what you stand for. In principle, this message is as old as business itself: Michael Porter described in his famous analysis of competitive strategy that as a company you can either pursue a low-cost strategy or a quality strategy. Doing both doesn't work. Entrepreneurs can't afford to be "stuck in the middle."[67] The value strategy that I am advising digitizing companies and IT manufacturers to follow here is, of course, clearly a quality strategy. For the goal of human progress, nothing should be too expensive. In any case, we should not imagine that the low-cost strategy really brings about progress for society.

Unfortunately, recent history shows that quality-oriented companies have a tough time in today's markets, which are heavily driven by cost-effectiveness and profit-seeking. About 50% of trading on the stock exchanges is now automated, often with the help of AI.[68] And the rating algorithms used are, of course, highly sensitive to corporate earnings reports. Since there are few people physically on the trading floors any more who could assess companies historically, holistically and on the basis of a philosophy, listed companies today are exposed to a merciless profit logic that is fully automated. If a young listed company wants to pursue a true value strategy and incurs high costs and low profits for a long time, just because it is the "right" thing to do, then it risks being punished by the capital markets. Furthermore, the capital markets cannot properly conceive of time. They live in real time, and a few months are already an age for impatient investors. Value-based innovation, however, takes time. "Good things come to those who wait" says an old proverb. And the famous maxim that "God is in the details" has been attributed to Mies von der Rohe, one of the most visionary architects of the 20th century. The tension between cost pressure and quality was also felt by Steve Jobs. "I want to make this a revolution, not an effort to squeeze out profits," railed Jobs to his CEO John Scully when it came to launching the first Mac.[69]

3.4.1 Economic Consequences of Automation

The tension between value creation, on the one hand, and cost pressure, on the other, is a core challenge for the humane and value-based progress that I propagate here. Nobel laureate Milton Friedman argued that company representatives are not responsible for value creation; their only goal must be to make as much money as possible. A move away from this economic rationale would, in his view, massively undermine the foundations of our free society. If we follow the Friedman doctrine, digitalization is more than welcome since it helps to massively reduce corporate cost. Digitalization accelerates automation; a trend that has been pursued ever since the beginning of industrialization. It can replace companies' expensive social systems with digital workflow systems – and now, as of recently, with AI. Not surprisingly, a wave of automation and AI transition has set in in almost all sectors.

This has already led to robot-based manufacturing in many companies. According to a statement by a VW union representative, the production lines for automobiles are already over 90% automated.[70] Amazon uses more than one million robots in its warehouses.[71] In China, there are logistics centers called "dark factories," so named because only robots work there.[72] No matter where we look, more and more supermarkets, call centers, train stations, airports – and even restaurants – are dismissing employees and putting machines in front of their customers.[73] Anyone who walks through the streets with their eyes open these days sees the world around us turning into a faceless self-service shop. And if customers are frustrated and hate to be treated like this, who is around to address their concerns? If they avoid the sack, the people still working in their respective industries are impacted on a day-to-day basis by the shifting sands of the technological age. Many, it seems, do not like these developments, and express this in their own way by no longer bonding with their employers. Younger people (aged 25–34) in the USA are changing jobs about every two-and-a-half years on average.[74] Some employed in manufacturing speak of "factories of fear" where they need to avoid robots or submit themselves to machine orders. If companies use digitalization solely to boost their profits, they are unlikely to contribute to human-centered and value-based progress in the 21st century.

That said, monetary value is of course also a kind of value. Money is the most efficient medium of exchange that we can imagine. It has become so important for today's economic thinking that some colleagues (especially in management schools) can think only of money or profits when they hear the word "value." For sure, companies need money to survive in daily business. They have to make profit reserves to secure investment. Automation can be critical to survival for some companies, especially if they rely on a low-cost strategy. Against this background, there is an ongoing debate about how firms could fairly share their profits from productivity gains with those who are being rationalized away. One recurring theme in this context is the provision of a lifelong unconditional basic income. Some political parties have started pushing for this idea. If the political and economic forces were to create a long-term sustainable financing of such a basic income, the curve of progress could have a positive inflection for some people. Especially those who have lost their jobs, it is argued, could then devote themselves to activities they really enjoy doing. But is that really so?

An interesting tweak to the scenario of peaceful automation is one I want to throw in: Friedman-Capitalism could be maintained, but banished to a standardized and homogeneous niche; namely, the market niche in which all the products are manufactured that hardly need any human quality work. This niche certainly includes IT products (computers and mobile devices), homogeneous durable goods (detergents, cars), and also services such as, for example, credit card services. Maximum process automation of these product and service areas would create free capacities in terms of labor and capital. But these must go somewhere if social order and peace is to be maintained in societies. So where could they go, if not to a few billionaires? One thought is that labor and capital could find entry into a second market niche where the "liberated" people funded in part by a basic income re-embrace more analog and manual manufacturing of products and services. Here, this investment could be driven by value qualities. The result would be fresh competition for those product and service areas that have already regressed as a result of overautomation (e.g., textile and food production). In other words, it is feasible to envision that people set free by automation could be re-integrated into those product and service industries that might specifically benefit from re-embracing manual labor. My idea might sound unrealistic at first. A look at the company Sonnentor shows, however, that such a re-allocation of labor is already happening. Sonnentor relies on manual work in the field of herb harvesting. It does not work with the available machinery, knowing that its competitive edge is the quality of the herbs, which are otherwise destroyed by automatic harvesting. Still, sophisticated machines are used by the company where human capabilities are suboptimal, such as when sorting the herbs with AI. Johannes Gutmann founded the company in the 1980s and had to accept his peers' accusation of being a "crank."[75] However, over the years, he was able to convince more and more Austrian farmers to once more pick up their tools and work manually to grow spices, herbs and grains, and to refrain from using machinery, fertilizers and chemical pesticides in the fields.[76] Today, the competitive position of the company surpasses that of its competitors by far: the value qualities of herb effectiveness, healing power, purity and quality outperform the market. Not surprisingly, the market success of Sonnentor speaks for itself: it now makes over € 60 million in sales each year, with healthy growth rates on a yearly basis.[77] The people I have met so far who work for Sonnentor are highly motivated because they all know that they are creating something of value and that they thereby have a purpose in their work life.

If we look at the Sonnentor example from a bird's-eye perspective, we could take it as a potentially symbolic case for a larger development. It becomes feasible to imagine how some of the people affected by today's progressive automation could be migrated to such re-manualized industries. Especially those people who are not able to effectively use the financial freedoms they gain from a basic income. Some elite theorists today gaze at their theoretical drawing boards and imagine that all people are equally at ease with using free time and money in optimum ways. This is, however, not necessarily the case because many people – as I will outline in chapter 8 – cannot really handle too much freedom. They become vagabonds. This is why I suggest integrating them into communities of re-manualized laborers for a few hours a week.

The idea of re-manualizing and re-socializing some industries does not mean that these companies and industries will be completely focused on manual work; instead, workers will only work manually in those areas where manual labor is the prerequisite for good value qualities in the end products; value qualities that have been lost to us in recent decades due to over-automation. That value qualities have been lost is confirmed by a brief look at today's realities: in many countries, it is simply taken for granted that food no longer tastes of anything, when a wool sweater sublimes into lint after being worn only twice, when ceramics are no longer properly worked out and woodwork has no patina. I make the argument that it is important to revive these value qualities in the interest of progress. It's about reviving innovation around traditional values, an area where Europe especially has been striving. Automated businesses that cannot compete with this traditional quality would probably lose their right to exist. From a bird's-eye perspective there would be two very different market niches: Niche 1 would be the segment for globally standardized products, which are manufactured in an automated and cost-efficient way. Niche 2 would be the area where manual and local production begins to flourish again, supported by state subsidies that redistribute the benefits of productivity gained in niche 1. If one looks at the herbs and spice market, one part of this path has already been beaten by our friends at Sonnentor, who were ranked third most popular brand for herbs and spices in Austria in 2024, with considerable market share.[78]

Of course, mainstream economists will ask incredulously how naive one must be to imagine that machines are being dismantled in entire industries and people are sent back to the field? I have no problem with accepting such criticism. From my point of view here, progress shows itself solely in terms of which value qualities are effectively created and which values are nurtured for people in the creation process. The common good is at the center of my thinking, not the maximization of shareholder value or any automation index or fictitious idea of progress through new features.

3.4.2 Ethically Questionable Workflows

Do we solve the labor market problem if we somehow integrate full automation with unconditional basic income accompanied by new market structures? Only partially, because the intermediate area, in which business processes cannot be *fully* but only *partially* automated, continues to be a challenge. If technology does not allow all process components to be digitalized, people will continue to be "harnessed" into digital workflows. They play the missing technical process piece, so to speak. FoodIS is a good example: orders, payments, invoices and courier assignments are all automatable, but the delivery is not. Delivery robots or drones are not expected in the foreseeable future in inner city areas for reasons of safety and noise pollution. And so the bike couriers are squeezed into an otherwise almost completely automated machine workflow, or digital "corset."

These machine corsets have become increasingly ungracious. The wages paid for such jobs are ridiculously low; accordingly, the people working in such roles have very little financial security. At Amazon Turk, an average hourly wage of $2 is paid.[79] On digital platforms like Deliveroo, Amazon Turk or Uber, people are competing for allocation to individual jobs or shifts; often they do not know when they'll get their next job. As a result, workers lose control over their time; they lose the freedom to self-determine their daily lives when they depend on this kind of job. And this is only one part of the problem. When people work in digital workflows, they are constantly tracked and monitored for how well they work within the machinery. While a cyclist at FoodIS can still choose which route to take for pizza delivery, his speed and punctuality are recorded. And whoever doesn't meet performance targets is relegated to a group of riders that are barred from competing for the most lucrative or desirable shifts.[80] From a value perspective, not only is there a lack of privacy but above all there is a permanent pressure to perform, which endangers the health and dignity of the workers, not to mention their freedom.

Further automation can even aggravate this. My students assumed that in the future, an AI agent would determine the delivery routes and guide the couriers accordingly. If that were to happen, the productivity and efficiency of the delivery would probably increase; yet, bicycle couriers would then also lose their last drop of autonomy. It is hard to speak of humane progress in such working conditions, which are more reminiscent of early industrial times than anything supposedly modern or progressive. Productivity and profit for shareholders that is predicated on a loss of value in terms of dignity and freedom for the employees can hardly be said to be a fair trade-off.

3.4.3 Value Hierarchies for the Right Value Priorities

Most business students today still learn that the most important value in business is profit or return on investment; that is, money. How could they learn to more systematically question this monetary value? Is it possible to judge digitalization projects rationally based on value creation for all involved stakeholders, where money is only one among several relevant values? I think this should be possible because rationally there are a number of criteria that determine whether one value is higher or lower than another. How to use these criteria to question the value of money can be seen with the help of a digitalization project undertaken at the company Disney.

Deep in the basement beneath the Disneyland Resort Hotel in California, laundry workers clean thousands of towels, sheets and blankets daily. The workers empty the dirty laundry into huge washing machines and extract the clean linen later in hot mangles. Until the year 2008, most laundry workers were quite satisfied, despite their hard physical work, and proud to work for a company that gives so much fun and joy to its visitors. Then, however, a shift occurred. In 2008, an electronic control system was introduced in the laundry facilities. Before the system's introduction, the washers had manually counted and written down the number of folded sheets, pillows and towels. Now this was taken over by the digital control system. A little time-saving on the part of the employees and higher precision in the counting initially promised a positive balance of values. However, the company decided to expand the system even further and turned it into something employees started calling their "electronic whip." The system didn't simply digitize the process but instead began to visibly display the productivity of each individual launderer, notably on large displays throughout the facilities. Each employee's name was assigned a color: if a launderer did not meet the managements' performance target, his or her name would be listed and marked yellow or even red. Similar to the scenario in Charlie Chaplin's movie *Modern Times*, laundry managers could now understand how well the performance goals of each employee were being met. And even the washing machines would give the employees feedback. If the worker failed to fill the laundry fast enough, a red warning sign now lit up to remind him or her to hurry up.[81]

In this example, material value ethics would consider the relative importance of the values created for all stakeholders. It is clear that the community of launderers suffered through the system's introduction. There is no longer camaraderie but competition among them. This undermines values such as solidarity and trust, albeit within the framework of a certain positive competitiveness, because the color system is communicated to employees as a "game." However, the pressure of monitoring leads competitive employees, who might otherwise enjoy the new system, to forgo bathroom breaks in pursuit of targets. While the male employees at Amazon logistics centers who find themselves in a similar situation pee into empty bottles to save toilet time,[82] this is a tad more difficult in the laundry setting. Consequently, the value of health is undermined, underlined by an additional increase in injuries. Employees are no longer free to pursue their work as their natural human rhythm dictates. They have lost the freedom and self-determination to work within their personal limits as they have done in the past. The employees do not call the system their "electronic whip" for nothing. Positive values such as community, joy, solidarity, trust, health, self-determination and dignity can hardly unfold. Only the value of competitiveness seems to flourish among high-performing employees, as well as of course the increased productivity and the monetary advantage to the owners of the company. Which values now weigh more heavily from the point of view of digital ethics?

Intuitively, many are likely to agree that the "higher" values of the launderers are so badly damaged that increased productivity, more money for shareholders and the competitiveness of some employees cannot make up for it. Those who intuitively feel this way are, as I will show below, ethically correct. But how exactly do you recognize the "higher" values compared to the "lower" ones? Scheler argues that the relative importance of a value unveils itself through our human intuition to *prefer* it.[83] This of course sounds dangerous. Because what if we prefer something that is actually bad? Can we really trust in our own intuition? Which objective rank order criteria can be used to actually educate our intuition and make it more reliable? In fact, Scheler himself suggested exactly such criteria, which I would now like to apply to the Disney case.

The first rank-order characteristic of a value is its *enduringness*. The competitiveness that was introduced by the system is certainly positive and also increases the pleasure of working at certain points, but only if you are physically and mentally fit. The value of competitiveness is not a permanent or enduring one. Competitiveness is countered by the value of a person's health, which can certainly suffer when one has to work consistently under high pressure every day. The lasting value of health is more enduring – according to Scheler's schema – and must therefore be ranked higher than short-term competitiveness. Also, monetary benefits are not permanent. While Disney Resorts may have increased profits when it first introduced the electronic whip, achieving potentially the same productivity with fewer employees, the question arises as to whether this profit is sustainable if employees' health and satisfaction deteriorates over time.

The second rank order criterion of a value is its intrinsic nature. Is a value like competitiveness, for instance, desirable for its own sake? According to Scheler a value is higher in rank the *less it is instrumental to* other values. Competitiveness, for instance, is not intrinsic, but instrumental to promoting something else; for instance, order, profit or power. Profit is also not a fundamental value, because it serves above all the satisfaction of investors or managers, and possibly also the security of jobs. But, as the saying goes, "money can't buy happiness!" Money is not an end – happiness is.

The third indicator of high values is that they are *less divisible*. Instead, they grow as they are shared, whilst low values shrink as they are divided. So, love or beauty cannot be divided but grow the more they are shared. The laundry community, as it operated until 2008, gave the employees a sense of unity: an indivisible value that grows the more one shares in it. The pride of working for Disney united them. In contrast, after introducing the whip, money incentives led to competition and division. The profit that was created is divisible. If the laundry employees' performance is tied to monetary incentives but at the same time the salary budget is fixed, employees compete for their wages or for their positions. This undermines any sense of community.

The fourth criterion of higher values is that they leave behind a *deeper satisfaction* than the lower ones. Thus, the values of community and pride that originally prevailed in the laundry gave the employees a purpose. Even for competitive employees who may surpass all performance goals, it is questionable whether they can draw as much satisfaction from their lonely excellence as they could have drawn from the pride and fellowship that had bound their colleagues together in the past. Even the managers, who may appreciate the increased profits and performance, may not experience such a deep satisfaction in exercising their function any more, because the sense of monetary reward cannot outweigh the loss of community and trust.

Finally, there are values that are of universal nature – such as goodness itself – that do not depend on any value-bearer. We respect their importance as such even when there is no context for them. Anyone who takes the Universal Declaration of Human Rights seriously would probably agree that the value of human dignity bears such an absolute nature. Regardless of the context – whether in war or in court – the dignity of a person is always worthy of protection, no matter how rich or poor he or she is. If, on the other hand, you consider a value like money, then it is clear that this is not the case. The working conditions in the laundry with the electronic whip are disgraceful when observed through this lens. The question, however, is whether today's society can train its ethical consciousness in such a way that it is even able to recognize this. Scheler already observed how difficult it is to correctly assess the rank order of values. He brilliantly wrote that in addition to the ranking of values that follow the stated criteria, there are also the norms innate in a given epoch. And in our current order, the values of money, productivity, efficiency and competitiveness are certainly the highest ones around. They are the golden calves of our time. They determine the "style" of our thinking and the sensitivity of our "morality." Unfortunately, today's style has little to do with ethics and the universal order of values that follow the criteria described.[84] The "order of the heart" is distorted in many.

3.5 Digital Ethics From an Economic Perspective

Against the background of what has been said so far, let me recap Fig. 2.2. When it comes to the first development of a product roadmap for FoodIS that did not focus on material value ethics, the balance of values looked initially promising at the individual, organizational and societal level. The satisfaction of many stakeholders increased. Households gained comfort because they did not need to cook and shop any more. New organizations like FoodIS created new jobs. Restaurants got additional orders, which increased their profits and also gave their employees more security.

At the height of these positive developments two paths emerge: The first is the path of humane and value-based progress, in which the potential of digitalization at all levels is used to add additional value for households, restaurants, cooks and couriers. That is what my students described when they followed Value-based Engineering. The individual benefits created for all stakeholders add up to a new form of "shared value" in society. The curve in Fig. 2.2 rises.

The second path is the one dictated by classical economics with its "value-free," technology-centered economic rationality. Here, the added value of digital investment for all those involved is not only exhausted at some point but negative externalities are created and regression sets in. How? Economic rationale would argue, for instance, that the level of pay for couriers should be based on the supply and demand for this kind of labor. This logic would recommend using digitalization to pay bicycle couriers as little as possible, i.e., by using automated auctions for job placement or automatically generated, flexible work contracts which can be terminated at any time. Microwork platforms like Amazon Turk, which work in this fashion, are contemporary examples.[85] Such treatment of couriers, however, creates negative externalities such as job insecurity, salary uncertainty and unhealthy competition between couriers. Furthermore, the profit-maximization logic will suggest efficiency-oriented route optimization through the couriers' app. Perhaps, even digital benchmarking between the couriers could trigger their physical exhaustion and long-term productivity loss, similar to the Disney Resorts laundry. There are many people who can work as bike couriers after all, and who are willing enough to accept such conditions because they are regarded not as humans but as replaceable "resources." Labor supply seems unlimited since no education is required for the job and almost everyone has a bike and a cell phone. Hence, it is rational from a classical economic point of view to develop the digital technologies used in the company in such a way that it minimizes labor costs for couriers and exploits productivity as described.

However, the fact that the employer considers the worker to be completely interchangeable and exploitable at all times will cause many couriers to become more unreliable, reduce their loyalty towards their "ad hoc employer," and possibly also lack politeness towards FoodIS customers. As so often in life, what one shouts into the woods is what will echo back. Pleasant contact with customers or a commitment to the employer seems out of the question in today's hardcore productivity scenario. What is more, as the students anticipated in their first roadmap homework, a feedback system is now required. Customers of delivery services must be able to complain about couriers as some may turn rude or unreliable. The company, FoodIS, has to invest in monitoring couriers at all times via GPS, so that no one makes trips that are of no use to the company. It must set up evaluation systems and introduce sanctioning processes or clarify legal scope in labor law. A costly negative culture of mutual surveillance and distrust is created, with the digital platform acting as a catalyst for all these negative externalities. Ultimately, there is a loss of added value on an individual, organizational and societal level due to increasing dissatisfaction, disloyal behavior, loss of freedom, rising health costs due to potential accidents and depression or the alienation of cooks and guests. Not all of these costs can be measured by economic indicators. But this does not mean that they do not exist. What is effectively happening is value destruction instead of value creation. There is regression instead of progress.

3.5.1 New Realism and the Necessity of Digital Ethics

Will companies be interested to hear that digitalization strategies that are not thought through lead to societal regress? In reading this chapter, the economically afflicted reader may have been beset by a big question mark. The ideas of the students are really great and important for society as a whole – but can the management of a company like FoodIS actually base their digitalization strategy on such a value system as Scheler proposes? Friendship between the couriers? Enjoyment of couriers while cycling? Freedom and privacy on the job? Allowance for more time between jobs? Even providing customers with a healthy dietary education that will undermine the ordering of high-margin pizzas? Such value-based service proposals are all investments in technical functionality that bring nothing to the company's bottom line. Even if only a few of these ideas were implemented, they would be costly and risky for the business. For example, if the couriers were to hang out idly in cafés and no longer take the quickest route to the customer? The modern manager would argue that companies are not charities. When it comes to management as it is taught today in business schools, there is no true concern about the long-term character changes, virtue effects or societal implications of technologies. "The business of business is business," as Milton Friedman famously said. And so business is not in the business of creating the true, the good or the beautiful that might possibly be engendered through AI or digitalization in general. Business is about job growth, margin development and competitiveness. That alone requires a manager's attention. This implies the need to keep the technical development costs as low as possible and the labor costs minimized. And it is against this harsh management reality in today's organizations that managers reading these lines will wonder whether it is not enough for employee motivation to invite all FoodIS couriers once a year to a fantastic retreat with free beer, rock music and a raffle? Is it not enough to show in other ways that you think socially? For example, by transferring 1% of your profits to some charity that might fund some nice bicycle trails in the Congo. Is such charity not enough when it comes to being ethical? After all, one stays (if possible) within legal boundaries as well; such laws might even provide bike couriers with a free cell phone change soon, as well as rainy-weather surcharges and maybe even a 10-min break between jobs (at least in Europe).

Frankly speaking, no, this prevailing business-thinking is no longer enough in the digital age. The idea that one can outsource digital ethics – if not corporate ethics as a whole – to a few feel-good activities, alienated from the organization's products and purposes, is not sufficient. Similarly, the idea that we can simply delegate ethics to the legal department, which then takes care of a minimum of justice and fairness in products, is also not enough. It doesn't suffice anymore to be a good man at home in private and not when you are out there in the company. Why is it not enough? Because digitalization has led to such an increase in the visibility and transparency of everyone and everything that unethical conduct or a lack of care is much more difficult to disguise – whether it be dark money accounts in Switzerland, dubious investment strategies in Panama, illegal diesel software in our cars, sexual harassment or systematic state surveillance. We live in a whistleblowing culture. And somehow, everything is revealed at some point, because everything we do leaves a digital trail, be it via e-mail, cell phone videos, short messages or even on already deleted databases that are dug up again by means of forensics. It has, therefore, become infinitely more difficult through digitalization to cover up even half-true or half-good behavior.

Added to this is the power of social media, where half-good or half-true is not enough, because everything that one says and does or does not do is potentially discussed in the public domain. If a company like FoodIS does not behave well towards its couriers, it does so for only a short while. In the long run it significantly damages its brand. Just think about the loss of brand value Uber incurred after it became public how its boss behaved towards an Uber driver and when it became transparent what brutal surveillance strategies the company was pursuing vis-à-vis its drivers and customers.[86] Digital information is like water; somehow truth seeps through on which values a company or an individual effectively represents.

It is also important to realize that it is not just companies that are being watched; we, also, as individuals are in focus. If I have a private affair as a politician, director, athlete or media star, I have to assume that it will eventually be publicly discussed on social media. Previously, there was collusion between politicians and the media to ensure that the negative behavior of public personalities (especially politicians) would be kept out of the limelight. Those days are over. At some point everything surfaces today on media such as Facebook, X or Instagram. And this is not only true for prominent people.

Through social media, we have moved into a "control society," as Deleuze once wrote. Here everyone is somehow watching everyone else and bad behavior leads to more gossip than ever before, putting potentially everything and everyone into the spotlight. This has good and bad consequences. On the positive side, I have the feeling that we live once more in a more sensitive political era. People are waking up from the sleepy 1980s and the roaring 1990s, and are discussing the micropolitics of everyday life again. They demand real change. They want to regain virtuousness and see their leaders behave. So, ethics and even religion become more important for everyone, both in their role as a private person and as an employee, as a service provider, entrepreneur or investor. The importance attached to the individual is the reason why in this book I often use very personal case studies and link values in our private lives with values in companies or in politics. In the digital age, where everyone and everything is closely networked, system thinking – market, politics, law and privacy – cannot be separated any more, let alone controlled. The path to the value system for the 21st century is through ourselves. Whoever walks it will change this world positively. Those who neglect it or do not have the courage to face it will not lead a happy life, but will instead be haunted and exposed.

4 On the Nature of the Digital Fabric

In order to reflect on value-based progress, it is not enough to reflect on a more value-based design of business models and technical requirements. We also have to deal with the digital itself, with the material properties of the digital fabric and the fundamental interactions of this fabric with the value equation of human life.

Usually, we know a lot about the substances we use and/or consume, for example, alcohol. To guide our choices in dealing with alcohol we have developed both habits of when, where and to what extent to consume it as well as how to signal its effects. Each bottle has an alcohol percentage label. In other words, we have a sophisticated system surrounding alcohol and this allows us to properly handle this seductive substance. We also know that alcohol is actually called ethanol and is a lightly flammable, spicy-smelling liver poison. We are well aware of the positive values of sociability, talkativeness and relaxation that are caused by alcohol. But most of us also know about its drawbacks: nausea, fatigue, loss of control, addiction and illness. It is precisely because of our awareness of the kind of value qualities set free by alcohol that we have been able to learn how to deal with it. We deal with it whilst largely preserving our free choices: in short, everyone can get drunk. However, alcohol is still recognized in our legal systems, such as in the Highway Code. And manufacturers of alcoholic beverages are required to mark their products with the respective alcoholic volume information.

But what do we actually know analogously about the digital and its "alcohol content?" The digital fabric equally influences our lives and values, such as our health, the community, our knowledge and our freedom. It does so at least as much as alcohol does. It pays, therefore, to ask which qualities or prerequisites the digital fabric brings with it as a material value-bearer, so that certain positive and negative qualities unfold for us humans in dealing with it. If we understood this a bit better we would be able to develop better digital technology, just as we now have really appealing low-alcohol alternatives. Furthermore, everyone would be more aware of how to use it well.

When I talk about "the digital fabric" in this chapter, I always mean the current silicon-chip-based IT systems (including AI systems), which consist of hardware and software components and run on a nonbiological substrate. Digital components of a system are usually networked via a high-bandwidth connection (at least in developed countries) and usually they are in an addressable standby mode; not only our smartphones, tablets and laptops, for example, but also future robots and software agents (like today's Alexa, Siri or ChatGPT).

 | https://doi.org/10.1515/9783111677736-004

4.1 On the Incompleteness of the Digital Fabric

In this chapter, I start by asking whether the digital fabric is actually capable of fully representing our human social reality and the world of values that defines us. Imagine you are single and you go to a local pub. You enter the premises and immediately perceive the good vibes of the place. At the bar you see a man or a woman and feel deeply attracted to them. You do not know exactly why, but you feel some kind of human connection. You see the beauty of the other person, his or her humor, intelligence, curiosity, openness, etc. In conversation, you discover more values than your curiosity can handle. Now compare the impression you have of this person with the following alternative: You are sitting at home using an online dating app like Tinder. Your swipe lands on the more or less photoshopped portrait of the same man or woman and you might even have access to his or her sophisticated profile. You read his or her favorite poem and information about personal interests and preferences. You have a good impression, are impressed by the beauty in the photos and immediately feel some kind of connection. How does this differ from the experience at the bar?

A sober comparison of the two types of perception, real versus digital, shows that experiencing a person in a physical encounter is more meaningful. The experience of values takes place intuitively, emotionally and physically. When I am physically present and encounter a human in a real way, then I can intensely experience the values that this person embodies because all my senses are in play. I feel physically attracted or repelled. Not only are my eyes involved in the sensory process but also the senses of smell, hearing, touch and even hormones act as sources of information. Objectively speaking, the real information input goes far beyond the alternative of just taking in a few megabytes of visual digital information.

However, the digital does not differ from the real through this sensorially more complete experience alone. Such a conclusion would be premature. Due to the complexity of our experience of reality, a real context also achieves a degree of intuitively developed completeness that digital reproductions cannot offer. The completeness of a real encounter between two people, for example, can be influenced by a certain radiance of the other person: a crooked eye, a dimple on the chin, moist skin – yes, by an almost infinite amount of inconspicuous details or stimuli. Not all of these details penetrate our consciousness. But whether it happens consciously or unconsciously, we associate certain value qualities with what we experience. The dimple on the chin of the person at the bar may remind us fondly of our father, or the woman's crooked eye of our first love. Many of these invaluable details will not be reflected in a digital system.[87] This may be because system designers are unaware of the importance of such details, which are different for each individual, or that each of these relevant details is so rare that an AI cannot recognize them as relevant, or alternatively, because some of these details, such as the aura of a human, are simply not measurable.[88]

Our real, biological world is characterized by an abundance of nonidentical repetitions, processes, encounters and situations. At the moment of conscious perception, these repetitions form for us a *singularitas* (unity) in which our memory of the past condenses with our expectations of the future into an unrepeatable experience of presence. One can describe this singularitas as the "uniqueness" of our individual experience in the here and now.[89] And it is at this very moment of singularitas that values, such as the value of sympathy, materialize. Of course, this is not only relevant in dating. In fact, we continuously perceive our entire environment and our fellow human beings in this singularitas. But when we understand perception in this detailed way, it becomes immediately clear why we humans have a far richer, wider-reaching and, above all, *different* kind of perception of reality than artificial intelligences.

If AI experts today suggest that a company's sales, customer service or sourcing should be digitalized and AI-supported as much as possible, then this could be bad advice. Companies should be aware that in digitalizing they do not do justice to human-enabled processes. If they disregard the singularitas reigning in the social ties between working people and believe that they can replace this experience with AI they may be deeply misled. The unique meeting of people in business and the values created in human networks of suppliers and customers yields positive outcomes that can translate to a big economic advantage. Just think of sales geniuses who arouse sympathy and trust among their customers in the course of a sales pitch; strategic sourcing experts who, through their quality awareness and accuracy in the act of purchase, intuitively choose the right raw materials, supplies and fair prices; call center agents who treat customers with friendliness and calmness when helping them; inventors who solve technical problems through their creative gut feeling, but also their curiosity; all of these human gems that lift up the social systems of companies with their warmth and gentleness. Without these human values, which only actualize in the singular perception of humans, no company can be successful.

4.1.1 The Big Data Illusion

Against this background, must we conclude that AI and digital services in general will never reach the degree of perfection needed to effectively replace social processes? Do we risk crude social and interpersonal regression if we disregard human singularitas and digitalize everything? Many transhumanists would vehemently disagree. They believe in the promise that more data automatically means more knowledge, and that machines simply need more data, more algorithms, and better ontologies to truly mirror situations and processes. Cambridge-based computer scientist Ross Anderson has described this belief as a typical "computer fallacy" because, he argues, it's not about having more data – but *the right* data to answer the right questions.[90] The philosopher and historian Yuval Noah Harari has coined the term "dataism"[91] for the belief that data is the key to the world. When "dataists" develop technologies or set up companies that they like to call "smart," the foremost goal is to collect as much data through their services as possible. They believe that services like dating portals, life-logging, or online advertising will work a lot better the more data about us users is available. They would probably argue that online dating is at least "efficient," ignoring that active online-daters often spend hours in the digital realm to organize a first date, which very frequently still goes awry exactly because the digital interface does not allow for transmitting the value qualities relevant in human encounters, and in dating specifically.[92] Instead of hanging out in a real bar, the digital generation wades through a mire of time-consuming hopes even though they all know its limitations by experience. So should dataists not concede that replacing real-world encounters with digital experiences is often inefficient? Probably they should, but they don't because they hold very deep faith: they worship the idea that the progress of technology – and specifically AI technology – will overcome current inefficiencies, and that it is only a matter of time until AI will similarly overcome its current value-alignment problem.

Dataists would argue as follows: Online dating sites could not only ask for much more information and perform more psycho-tests but also integrate things like a digital scent sample; when all is said and done, isn't the odor of a prospective partner important? First, scientific prototypes for sending scent samples already exist.[93] The smell could be conveyed in advance to the potential partner.[94] After all, scientific studies show that odor is a surprisingly important factor in dating. Or how about a glimpse into the apartment of the potential partner? Smart TV manufacturers have already equipped their screens with active cameras on many models to collect such data – why not use this information in dating services, or, for privacy reasons, at least a benchmark-index of how neat someone is?[95] Something similar is, by the way, already being done today when it comes to the private smartphone data of users. Some start-up companies have offered to analyze how well smartphone users look after their phones – whether they regularly recharge the battery in time, for instance. They then draw conclusions from this behavioral data regarding the character of the users, their tidiness, trustworthiness, etc. This information is subsequently regarded as relevant and can be used for decisions about whether that person should receive a loan or an invitation to a job interview, for example. As we can see, profiteering from and upscaling of our personal data is absolutely normal.[96] So, it follows that it would only be a small step to have our apartments examined as well. Dating portals could definitely benefit from better success rates for the first date if the tidiness and style of a prospective date's home were known beforehand.[97]

Quite apart from the fact that I personally observe such a potential loss of privacy and freedom with a shudder, I simply do not believe that the world is as simple as dataists imagine it to be. Data cannot create a "mirror of the world" as Silicon Valley entrepreneurs or some authors like to claim, nor true "digital twins."[98] Why can data not create reliable mirrors? Let's discuss the singularitas argument for the FoodIS story from the previous chapter: Ben – a fictitious FoodIS courier – is puzzled by the sympathy and friendliness he perceives when Mona Lisa – a fictitious customer – opens the door to receive the pizza she ordered. Somehow Mona Lisa's dark chestnut hair reminds him of his mother's, and she has the same beautiful blue eyes. Ben quickly casts a glance beyond her into her apartment and he can see that it is stylish, neat and homely. Mona Lisa opens the box, and a tantalizing smell envelopes the two, reminding them both of their childhood. Smiling, she gives Ben a generous tip, and he happily turns to his next assignment. Which value qualities drive this real-world situation? Sympathy, friendliness, style, order, comfort, joy, generosity – perhaps, even love? All of these are ultimately invisible value qualities unfolding between two people and things in a real context. No IT system can grasp this complex reality.

Dataists are therefore subject to a very strange control illusion. They believe that with their AI-driven big data models they can digitally grasp and understand our entire world and our lives by simply incorporating a few sensors into our homes. In reality, however, the things that really matter in life cannot be grasped more than perfunctorily by machines because values are only perceptible to humans involved in a particular situation. Values are not visible and cannot be measured digitally. Dataists might of course hope that this is not true. They might say that it is technically possible to install video cameras throughout the home and oblige every citizen to wear smart watches and sensors all over their bodies. The video cameras in the apartment could record Ben and Mona Lisa's smile; facial recognition software would pick up the sympathy between the two, and perhaps even arousal if skin conductance data was added. The video cameras could also record the flat and use an algorithm to figure out if it is tidy. The images of the apartment could be compared to a database in which all furniture is stored to determine how expensive and how new they are. On this basis, then, perhaps the "style" could be defined. If there is a fireplace, the IT system could automatically add the classification "cozy." And finally, the payment app could determine if the tip was above average and consequently deduce that it must have been generosity or satisfaction with the courier. Everything seems to fit neatly into a model!

The only question is: why did Mona Lisa smile? Did she smile because she liked Ben as a man? And was Ben surprised because he found Mona Lisa attractive? Or was Mona Lisa looking forward to an afternoon picnic on that very day? Did she perhaps think of her mother, who had told her the previous evening that they would go on an outing next month that Mona Lisa had long dreamed of? Perhaps Mona Lisa's smile was not for Ben. Past and future, spatially close or distant – nothing of what really determines our consciousness in a given moment is detectable by machines. Furthermore, one might wonder whether we can sensibly grasp the "style" of an apartment in a technical model. Isn't every style unique? Is a fireplace an accurate gauge of whether a home is "stylish" or "cozy?" And is it possible to automatically deduce "generosity" or customer satisfaction from an above-average tip alone? Maybe it was indeed the affection between them?

Machines only depict reality secondhand in an indirect, downstream and incomplete data model that tries to reproduce the truth but cannot grasp the world of values. The completeness of the real singularitas in which we live is not accessible to machines. And even the sensory information that is physically perceived, which materializes as feelings in living beings, is neither observable nor reproducible in its infinite complexity by a machine. Does this mean that the ability of a machine to participate meaningfully in the human world is colossally limited? Can AI ever be truly intelligent in a human sense? On the basis of which data should that be possible? How are humanoid robots supposed to behave like humans if they are based on such suboptimal datasets? Despite these fundamental and completely open questions, an unshakeable belief in AI is widespread not only among laymen but also among some world-leading "experts." This quasi-religious belief is so strong that some are even seriously calling for more use of AI in the legal world; a trend already commonplace in the USA.[99] Some "experts" claim to be able to prove that AI can judge people better than their own parents,[100] which leads others to suggest that one could have jobseekers speak with an AI first to determine whether someone is fit for a job. I call this phenomenon of exaggerated belief in the completeness of the digital fabric "the big data illusion."

The consequence of the big data illusion is huge losses of value for companies and governments at all levels. Companies irreversibly lose the trust of their customers when the latter learn about the scale of surveillance, and customer respect and loyalty are forfeited when they are forced to deal with AI-chatbots in call centers rather than company representatives. And the negative repercussions will not be limited to the anger and despair of customers shouting at call center bots. If competitive markets are still available, then customers will, in the medium term, end up looking for more customer-friendly and authentic, human-employing companies. When it comes to recruiting, businesses are also destroying the curiosity and excitement that university graduates might once have felt in applying to them, providing them only with an AI to talk to on their first interview. Governments lose the last bit of citizen respect when they allow machines in the jurisdiction. People would have to be very jaded and desperate to react differently. But even if they eventually accept the loss of their privacy and dignity with a shrug – which is what I expect – then this is associated with a huge social regression; a regression that even George Orwell could only have dreamed of.

4.1.2 Shadowboxing With the Absent

The limitations and incompleteness of the digital doesn't only have negative economic consequences: we also lose from a communitarian and personal perspective. This is because humans' imagination in the face of limited digital signals leads them to fall into what psychologists call "autocompletion." Do you recall a situation where you got a phone message that was too short or ill-formulated to be correctly and fully understood, perhaps a message from your boss or a friend? Or, have you ever been annoyed about the fact that you only got five likes or retweets when you expected at least 30? If so, then you know that sometimes you cannot concentrate for hours because your mind is awash with imagination about what your friend or boss might have meant in his or her message or why your popularity might have dropped on social media. Your friend is ghosting you . . . Maybe she doesn't like you anymore? You have no way of knowing because the digital medium gives you only incomplete information. So you invent stories about what the others might have thought and construed and you start spinning in circles.[101]

However, when you meet your boss or friends later, the ambiguity is often resolved in no time. The other person wanted to say something completely different than what you interpreted from his or her message. And it was the Facebook algorithm that decided it would not display your post to anyone in your circle of friends. When faced with incompleteness or remote untransparent AI decisions, we are easily subject to illusions and seek a deeper meaning where there is none. In the absence of a real human counterpart or context, we project our personal patterns of thinking and wishful ideas onto the information received. We go into autocompletion mode, and if the "autocomplete" we chose is negative, then we start what I call the exhausting process of "shadowboxing with the absent."

In online dating, this autocompletion behavior is easy to understand.[102] After several hours of information exchange per week, 90% of people report that they like the potential partner and see similarities with him or her before the first meeting. The information exchanged, such as career success, good looks, popularity, interesting hobbies and language ability, as well as the first messages shared are so convincing that the first date is a highly attractive prospect. After this first encounter, however, only 36% still like the person they just met in the flesh. The digitally raised impression that they could have "similarities" has in most cases vanished. It has nothing to do with reality. It was only a fantasy.[103] And yet, one of the core arguments for online dating is efficiency![104]

Are we fooling ourselves then? Why do we humans chase shadows? Science shows that, in fact, autocompletion behavior is normal for humans. It starts at a physiological level: there is a blind spot on our retina where we definitively cannot see anything. Nevertheless, we imagine that we have everything in our field of vision. We do so because our brain imperceptibly "completes" or fills in this blind spot by itself. In addition to this physiological phenomenon, psychological research shows how vitally important control is for us humans – and thence stems a natural desire for the complete understanding of things around us.[105] Behavioral research distinguishes three types of control: information control, decision control and behavioral control. Information control means knowing what's going on around us.[106] Decision control means having ways to react to this information. And behavioral control means that we get feedback from our environment that our decisions have been respected. For example, my desire for behavioral control is satisfied if my counterpart nods or smiles in agreement. With regard to the machine world, our behavioral control is satisfied, for example, when the automatic vehicle control in our car reacts immediately to our commands or signals to us with a flashing icon that it has followed our instructions. However, if the lamp lights up because of something we do not understand, then we imagine all sorts of reasons as to why it might have done that. In short: we seek control of our environment. It is only through control that we feel comfortable. When the desire for control is not met, people become angry and reactive.[107] Negative consequences ensue, such as loss of trust, helplessness, and reduced self-esteem.[108]

4.2 On the Lack of Integrity in the Digital Fabric

We have seen so far that the digital fabric can never have the same completeness as the real world because it cannot grasp values and the singularitas that give our lives their meaning. In fact, I would even argue that the very "DNA" of the digital fabric already determines this incompleteness, because by its nature it arises from division and not from wholeness:[109] it is made of bits.[110] Bits are symbols that stand for 0 or 1 and form the physical origin of all digital processing. Bits are combined into bytes, kilobytes, megabytes, and gigabytes. They integrate a rich stream of information that gives us perceivers an illusion of unity, especially in two-dimensional media such as high-resolution photos or videotelephony. But the primal technical character of any digital representation is a 0/1 split, and this split is passed on to all product derivatives. Perhaps it helps to imagine what flour feels like when you sift through it before baking. The consistency of the sifted flour is simply different than when it comes straight from the pack. You can use both substances. Sifted flour is often even important if you want to bake in a "fine" way, but it's not exactly the same thing.

Taken together, we can see that the digital, through its sifted bit-nature, seems more ordered than its analog counterpart. The bits and bytes are neatly sorted together into tokens, attributes, objects, functions, relations and sequences, resulting in an orderly looking puzzle of larger contexts. Digital processes in companies often seem more professional than those done by people. Neatly sorted personal profiles of individuals seem more structured than talking about the people they represent. Digital prints are easier to read than handwritten manuscripts. Due to its ordered, sifted 0/1 nature, the digital brings to the world value qualities such as order, professionalism, clarity, concentration and efficiency, which are very pleasing to us humans.

However, what is often lost – and we must be aware of this – is the possibility of grasping form and content in a reliable and complete way. The digital form is always well-ordered and professional. But the underlying content varies greatly. The handwriting of a student allows me to infer something about his or her character. Some write very small and neatly, and this may be reflected in their precise and accurately formulated content. Others write with huge letters and wide intervals; they often think big, but may lack some depth. Due to some consistency between content and form in a handwritten manuscript, I can get a corresponding idea of the person behind it. In digital works, however, this unified perception of content and form is lost. I get digital files that promise great professionalism because they are so nicely formatted, but when I delve deeper into the content, I often notice that it is not as clear and precise as the digital appearance promised it to be. The proficient-looking digital fabric can produce a surprising discrepancy between beautiful form and mediocre content, between expectation and reality. With AI chatbots this problem is even accelerated, because we are led to confuse extremely well-formulated letters, seminar papers or even wedding speeches with the person behind them, who may be far less eloquent. Negative values such as disappointment and dissatisfaction arise when AI hence stokes false expectations.

The discrepancy between the digital surface, on the one hand, and an often incompatible reality, on the other, is today encountered in many applications of the digital. In the realm of social media this discrepancy has become a gigantic problem because fake news looks just as professional – and thus, credible – as real news. For example, many Catholic voters before the 2016 presidential election in the USA received real-looking articles in which it was reported that Pope Francis would, if he could, vote for Trump. They believed in the perfectly presented format of online news; but, in fact, the content was fake. Fake news is abundant in social media and risks being spread through modern AI chatbots and personal assistants even more. Just consider that since 2021, Facebook has taken down around 800 million to 1.2 billion virtual bots (false identities) from its network *per quarter*! These so-called "bots" are computer programs that look like actual human Facebook users and automatically distribute political propaganda messages, among other things.[111] The digital form of these bots is so professional that most people are not able to recognize their artificial identity.

In other words, the well-ordered form of the digital can confound a realistic and true experience of the world. It promises an order and unity where potentially none exists. This inevitably leads to confusion and misjudgments, which, in the long term, give rise to negative values such as mistrust, insecurity and suspicion.

4.3 On the Error-Proneness of the Digital Fabric

The incomplete and sifted nature of the digital fabric is joined by a third challenge: the error-proneness of software systems. In his book *How Real Is Real? Confusion, Disinformation, Communication*, Paul Watzlawick described an experiment in which two subjects physically separated from each other get to see images of tissue cells from a supposedly medical diagnosis software program. The subjects are asked to learn to recognize the conditions of "ill" or "healthy" cells with the help of the software. Like many of today's computer programs, the system looks very professional; nobody knows how it works or if it even works properly at all. The experimental subjects received positive feedback from the computer system if they were diagnosing correctly and they got a warning if they made a mistake. In this way, they learned to distinguish ill and healthy cells through this feedback mechanism. After a certain period in which they performed multiple diagnoses, the participants in the experiment had to get together and explain to each other how to recognize the health of the tissue cells. What the participants did not know was that one of the two computers regularly gave incorrect feedback, causing the affected subjects to not properly learn to diagnose the health of the tissue cells. But instead of noticing the computer error, the misled subjects trusted that their machine must of course be right. Instead of questioning the confused computer feedback, they came up with very complex explanations for the (in reality false) diagnosis of tissue cells; they filled (autocompleted) the incomprehensible reality with their imagination. Worse yet: when the erring experimental subjects met with their correctly-trained counterparts and both had to agree on how to correctly analyze tissue cells, then almost always those with the wrong and complex explanations convinced the other party. Faulty computer programs not only stimulate people's imagination but they also lead to a shared embracing of false conclusions.

To be honest, I don't want to know how often similar situations occur in companies today. The problem, put simply, is that digital software-based systems are error-prone. It is true that machines make far fewer mistakes than humans in many areas: they are more precise, see more details, can repeat without errors and have no fatigue faults. But despite all these benefits, they make plenty of software-specific mistakes: the code may have errors; the processed data may contain errors (being outdated or containing duplicates); the system may not have sufficient data, or is being used in contexts for which it was never developed; the system sees bit-flips in messaging; or a hacker may even have subtly changed the program's data integrity.

Due to our long heritage of successfully using mechanical machines, we have gotten used to the expectation that these are working reliably and largely in an error-free fashion unless they are broken. We therefore overtrust that this is equally true for software systems. But that is not so. In fact, a large part of software development is spent on debugging. For example, in aviation, software engineers typically spend only 3 out of 24 months writing new program code. The rest is used for testing, debugging or documentation.[112] The result of debugging is an official error rate, a quality index for software. When systems are developed for high-security areas, such as hospital systems, manufacturers aim to get fewer than 0.5 errors per 1,000 lines of code. But the reality for any normal system is approximately 6 errors for 1,000 lines of code according to current statistics.[113] Of course there are also methods, such as those employed by model-based software design, that aim at lower error rates. But especially with AI-supported programming, the risk is that system errors replicate.[114] Complex systems contain millions of lines of code. A highly digitalized car, for instance, can contain up to 100 million lines of code.[115] In this case, 0.5 errors per 1,000 lines of code translates to some 50,000 errors. Doesn't it seem rational, therefore, that trust in such a highly automated car should be limited?

Against this backdrop, it may also come as little surprise that software bugs are increasingly causing serious problems. When aviation security systems fail then whole fleets of planes need to stay on the ground. Toyota was forced to recall 9 million vehicles for maintenance and had to pay $3 billion in compensation after software in their car brakes failed.[116] Experts fear that hundreds of hospital patients in the UK are falling victim to software errors every year.[117] In 2021 the Rutte cabinet in the Netherlands had to step down from government after a software issue caused 26,000–70,000 parents to be wrongly accused of making fraudulent child-benefit claims, resulting in demands to repay the (presumed and faulty) allowances.[118] In many cases, this sum amounted to tens of thousands of euros, driving families into severe financial hardship and resulting in over 1,100 children being taken into care between 2015 and 2020.[119] In the UK, meanwhile, 983 sub-postmasters were wrongly suspected of having stolen money from post office bank accounts and had to fight for almost two decades to defend their innocence. The UK legislator's final 2024 report on the matter summarizes the misery that this one system error caused:[120]

> From 1996, the Post Office piloted new Horizon software in some of its branches. This software was introduced across the network of branches in 2000. The software had faults which meant that some postmasters' accounts showed false shortfalls. The Post Office obliged postmasters to "repay" these false shortfalls. Some postmasters and others were suspended (generally without pay) and/or dismissed. Others were prosecuted for dishonesty offences (by the Post Office itself or by the Crown Prosecution Service (CPS)); some were convicted and imprisoned. Some were made bankrupt (in some cases on the petition of the Post Office: there are allegations that this was done maliciously). Some lost their homes. Some suffered mental or physical health problems as a result of their treatment or of the financial consequences. Some were harried as thieves by their local communities. Some suffered breakdowns in relationships with their partners, children or other families and friends. Several died by suicide.

Confronted with such critical issues of software use, dataists and the IT industry argue that human error has also always been present in all areas. After all, the only thing that counts is that software makes fewer errors than people. That's right. But luckily, we know all too well about human errors that are singular and don't replicate across a sometimes nationwide infrastructure. Human errors are captured by corporate or administrative processes that have been built to be fault-tolerant, thereby cushioning against them. In a dataist world such precaution is rare: due to our historical experience with largely reliable machines, most users and buyers of digital systems still trust that these are equally flawless; after all, this justifies our investment in them in the first place. If digital systems then produce unexpected errors, buyers and sellers tend to justify these somehow. Sometimes they simply do not *want* to see them, and if they do search for the source of the error then they often seek to lay blame on the system's human users. They look for the human error instead of the software error, and devote themselves to justifying even the machine's most bizarre results. What is happening here resembles Paul Watzlawick's experiment described above. The people involved, facing error-prone machines, come to sophisticated but wrong conclusions and spread these further. This then results in negative consequences for companies, such as confusion, disorientation, demotivation, and loss of control, which employees can only escape by switching to "autopilot" and no longer caring about the company and its IT mistakes.

The consequences for both governments and companies are severe. Companies lose substantial employee loyalty and motivation. Wrongly treated and misclassified citizens lose confidence in their government. Everyone has to live in uncertainty, knowing that they could be wrongly classified any day. Bizarre, Kafkaesque scenarios are no longer a thing of fiction. Instead, the errors leave societies increasingly in a state of retrograde chaos.

Against the background of these developments, a lot of hope is placed on software developers' responsibility, their diligence, and the maturity of their development tools. Certainly, all of us who work in the IT world expect that industry advancement will rule out or at least mitigate most of the horror scenarios I described above. Maybe software could write itself at some point in the future and use even more artificial intelligence? However, this trust in the "principle of responsibility" of technical development departments might be misplaced. Not even the developers and administrators of IT systems know exactly how their machines work in detail. Many people are involved in the development of a single system. Where artificial intelligence is used, opaque black boxes produce behavior that is by default not precisely explainable.[121] Similar to users, developers project their personal explanations onto their inventions. They often assume that they have it right without really being able to prove it. Sometimes they "tune" their systems a bit, if certain details do not fit. However, many of these explanations and tunings are not documented or verifiable, especially when IT departments are not given enough time to understand and document what their systems are doing or where problems are dormant.[122] In an article entitled "Professional and Ethical Dilemmas in Software Engineering," longtime Siemens engineer Brian Berenbach opines that[123]

> we see a rush towards system delivery that can lead to problematic ethical behavior, such as the rushing of engineering jobs, the difficulty to honor agreements, the lack of comprehensive and thorough evaluation of computer systems, a promotion of "fictionware" and the tendency to sweep a lack of quality under the rug.

In summary, it must be noted that reality distortion is not only the scourge of ordinary system users and buyers. Digitalization projects must therefore manage to ensure a much higher internal and external transparency of their technical details.[124] Information control, decision control and behavioral control should be the norm at all system levels so that we can trust that the values of quality, safety and care are present. Digital systems must be structured in such a way that the modules which they consist of are so simple that they can be replicated by changing development teams. Errors must be made transparent so that the degree of potential distortion of reality is always clear. There should be analogous backup processes for all digital processes by which people are affected. "Good" digitalization does not always mean creating more powerful and faster machines with more and more data. Rather, it means sophistication under the condition of entropy minimization. It means creating value without immediately undermining it elsewhere in a second step due to software errors and other flaws. There is a difference between true progress and more technological change that is too often a step backward.

4.4 The Digital Has a Life of its Own

Two of my favorite technologies show how simple, up-to-date devices can enrich our lives directly, helping us unleash some positive values without simultaneously undermining others. My first favorite device is the original (not Bluetooth-based) Bose noise-canceling device, which filters away all sounds in my environment, leaving only that which I wish to hear. The headphones allow me, especially during air travel, to protect my health, find peace on the plane and increase my enjoyment of music and audiobooks. My second favorite technology is the lumbar extension machine available in Kieser fitness studios, which is designed to strengthen back muscles. It has helped me to regain my upper-body strength after years of sitting at a computer for hours at a time.

What I particularly appreciate about these two favorite technologies is that they are always patiently waiting for me. What I mean by this is that they never demand that I use them. They are silent. My Bose device does not flash at me irritably. And the lumbar machine is in the Kieser studio. I never feel obliged to use these devices, nor do I depend on them. And that means I have room and time to be undisturbed. Why is that so important? We humans need free space around us into which we can unfold. By that I mean not only physical space but "room to breathe." When people invade our private space, even with the best of intentions, they often "get on our nerves." As described above, people need control over themselves and their environment. They want to be self-determined. And if you deprive them of this freedom, it can make them not only depressed but also aggressive.

Accordingly, this means that we want to enter into an exchange with the world in a self-determined way. We want to interact with it unhindered. And in this freedom of dealing with things, the environment and others, we experience ourselves and can be happy. We are happy when we experience ourselves as value-bearers. When I decide to go back to the gym after a long time and train my back there on the lumbar machine, I have a sense of accomplishment. Not only did I show discipline by going there in the first place, I also feel my body, the pleasant pain of effort and the comforting feeling in my limbs that sometimes lingers for days after. I also feel like I'm getting stronger if I do these exercises regularly. Philosophically speaking, Scheler once wrote: "The person is a continuous actuality. The person experiences virtue in the mode of the 'being-able-to' of this actuality in regard to something that 'ought' to be done."[125] In short, if I give myself the time and space to develop my health and virtues, then I strengthen myself as a value-bearer.

Many of our digital devices and services today deprive us of this kind of mental and physical experience. Through their own "machine lives" they influence the way we live in the world. In 2025, people worldwide spent on average 4 h and 37 min per day on their phones, which adds up to roughly a day per week or six days per month.[126] Our smartphone is like a lively friend – and becomes even more so as we increasingly speak to it through our conversational agent technologies (e.g., Siri, or favorite bots configured on character.ai). We no longer wish to be without the device or software agent. And these systems insistently prod us to invest our lives in activities we would not have chosen without them. For instance, spending over two hours on social media a day.[127] Why do we do this? What is going on here? How can it be that the digital, smartphones and social networks, messengers and e-mail services, as well as news sites and video platforms, virtual game worlds and music streaming services compel us to return to them again and again?[128] The simple answer is that they are not really "lifeless." They have a barely controllable machine life of their own that permanently absorbs our attention and is deliberately designed to pull us in.[129] In the Western world we believe that a machine is "neutral" because it is of a material nature.[130] But is that true? Aren't we getting really attached to our AI's?

4.4.1 Business Models for Addiction

Another example – this time Amazon Kindle – shows that there is no neutral connection between the machine life of our devices, on the one hand, and the digital market economy, on the other. Today's digital economy is (still) largely financed by advertising. And, therefore, it is entirely predicated on the fact that we pay attention to what is being offered online: services, products and news. Our human nature forces us to pay attention to everything that changes noticeably in our environment; for instance, an ad appearing on the Kindle in our vision space, or a moving video announcement right next to a YouTube video we are watching.

In psychology, this is called "Orientation Reaction."[131] We orient ourselves in the world in reaction to any stimuli that might be relevant (originally to aid survival). This attention, however, is a scarce human resource and the advertising industry has contended with the fact that the attention span of us users has dropped so much in recent years that it is becoming increasingly difficult to justify the price of advertising space. Gloria Mark, emeritus professor of human-computer interaction at the University of California in Irvine and author of the book *Attention Span*, found evidence of how drastically our ability to focus is waning. In 2004, her team of researchers found the average attention span on any screen to be 2.5 min. In 2012, it was 75 seconds. Six years ago, in 2020, it was down as low as 47 s.[132]

An impressive quote on what this actually means is offered by Ofcom's Children's Media Lives Project, quoting 13-year-old Suzy:[133]

> I find it really hard to watch movies. . . . So normally if I'm watching something on Netflix I get really bored of it and I just go on my phone and I don't even realise it, I'm just scrolling on TikTok then I look at the time and I'm like, "what the hell, I need to go to bed." My head of year said something about it the other day. He said it's because we all watch TikTok. Like our brains aren't used to watching really long things. That our brains aren't stimulated enough by it or something. I actually think that's true, because I can watch TikTok for hours and hours and hours and not get bored. But as soon as I put on a movie I get bored, and then just go back on my phone.

As a result of these developments, all those who offer advertising space are competing for the ounce of attention that is left. For money to be earned, it must be credibly argued by advertisers that we, the users, have actually seen the ads. And the longer, more frequently, and more intensively someone uses a service, the more likely it is that he or she has seen them. Therefore, the frequency, length and intensity of our online stay must be maximized. It is this market momentum of addiction maximization that leads service providers to endow our devices with a proactive and dopamine-manipulating machine-life form that pulls us in.

How does this digital addiction work? Why do we sometimes spend hours, even days, weeks and months on our digital devices and in virtual worlds? It's quite simple: the art of digital companies in perfecting so-called "Addictive Designs" is very mature.[134] A few simple examples illustrate this: Let's take, for example, the phenomenon of video platforms such as YouTube or Netflix automatically queueing up the next movie as soon as one is finished. Or that on Facebook and LinkedIn, after accepting a few contact invitations, we can directly see the next "people we may know." This automatic refilling of content without interruption is done very consciously because platform owners know that it leads to a barely controllable overconsumption. An analogous study on this phenomenon has been done by Professor Brian Wansink at Cornell University. In the area of food consumption, he had people eat from a soup bowl that was automatically refilled from below without the subjects noticing. When people ate from this bottomless bowl, they consumed 73% more calories than the peer group who ate from a normal bowl.[135] Our Internet services are like bottomless bowls.

Another experiment well known in the digital industry is called the "pigeon test." In the 1970s, the psychologist Michael Zeiler investigated the phenomenon of rewards.[136] He used pigeons that were kept in a cage, teaching them that they would get one grain of food in regular 3-min intervals. They would reliably receive the grain when pecking at a particular switch in their cage. Another group of pigeons was not granted this regularity, and did not get the food at these fixed intervals. Sometimes it worked: they picked and got grains. Sometimes it did not. The consequence of this unpredictable feedback from the food apparatus was that the pigeons pecked twice as often at the switch as the pigeons in the reliably fed comparison group did. Pigeons peck twice as often to make sure they at least get something to eat. Michael Zeiler concluded that unpredictable feedback in pigeons means that they will double their effort to get the same reward, and it is precisely this mechanism that is being exploited today by digital operators to make people return to online services. A good example of this is the Like button on Facebook. Since users never know how often their posts will be liked, but appreciate this little emotional reward, they are always tempted to go back to the platform to check it out. The same goes for social media like X, where you want to know how many likes and retweets you got. You want to know if important people have responded to your messages, how many comments there were on your blog and what happened in your online game while offline. The unpredictability of the many small rewards of the Web makes us feel constantly tempted to return to it and click like the pigeons.

It does not have to be that way. Online services do not need Like buttons. If they really wanted to use them, a short message on our phones could show us in aggregated form every few hours how many there were and from whom they came. The digital service would still have a life of its own but in a slimmed-down form. In addition, the rate of change that occurs today in real time on platforms could be reduced. For example, Facebook, Instagram etc. could publish the likes or hearts according to a fixed timetable, so that every user knows that, for example, the aggregate statistics will only be shown once a day; e.g. at 6 p.m. As a result, users would feel less tempted to constantly go back to the platform to check. These are very small technical changes that could reduce the addiction of users massively. "Small," of course, is a big word. A platform like Facebook or Instagram, which earns its money on the back of users returning and being hooked, would need to rethink their entire business model. This would be a huge step indeed.

4.4.2 The Problem of Self-Interruption

The constant change of digital surfaces is a constant interruption for us users. If we sit with friends at dinner and the phone is lying next to the plate or vibrates to alert us to a particular message, then the table conversation is over. When we are reading or writing text on the computer and suddenly a small window pops up with an instant message or e-mail, it interrupts our productive creative work. In 2025, an office employee could expect to receive an average of 117 e-mails per day.[137] Even though many of these e-mails may consist of advertisements and, therefore, are not really read properly, this still constitutes a considerable flood of e-mails, and this barrage is only added to by the numerous short messages we receive. In short, the number of messages we receive every hour leads to a permanent interruption mode.

Scientists have shown that this mode of being constantly interrupted leads to the habit of self-interruption.[138] Even if our phone does not show a new message and is next to us, we activate it on average 205 times a day.[139] If this staggering number is correct, and we estimate 18 hours of wakefulness per day, then this would mean that we roughly check for news every 5 min and thereby self-interrupt our mental and working flow. Self-interruptions of this kind are extremely unhealthy because it normally takes us humans 15 minutes to achieve a flow state in a primary task. Flow is the mental state in which one is so engrossed in one thing that one forgets everything around oneself and becomes truly productive and also happy. Mihály Csíkszentmihályi investigated this state, in which one becomes one with oneself and the world, in more detail; the phase in which the student learns, the scientist writes his book, the artist paints his picture, the engineer builds his machine and the athlete outgrows himself in competition.[140] It is the flow in which values can best unfold: knowledge, creativity, beauty and perfection.

However, our digital devices impede this flow. This is because someone (often we ourselves) activates them over and over again. And at that moment, the flow that may be just about to materialize dries up and the flow of value collapses. What is happening? Our inner thumbs on our smartphone and the shimmering surface of all our lively devices turn us into purely reactionary beings without inner leisure and rest; into beings who have no more opportunity for self-development but are pushed into a strenuous passivity in which there is only one goal: to function by response.

I want to emphasize the issue of self-development or flourishing here: every person needs a space in which to unfold; either alone or together with others. We speak of "free space" for ourselves. We build ourselves spaces in which we can withdraw and in which we extend and "flow" to be and work as we wish. Or we sit down at a table in a dedicated room (a kitchen, a dining room, a bar) to share the flow of events with others. These material and immaterial spaces support our individual or common development. In the past we have been able to develop undisturbed in these rooms. Now, however, our smartphone is on the table or our personal agent is with us in the dining room, listening into everything we have to say. As a result, our utterly private cocoon of self-development is disturbed. Our human space is occupied by something else. We are no longer in full control. We can no longer freely and independently become one with the world and develop ourselves creatively because we are interrupted and disturbed by the networked devices with their commercially manipulated machine life. At least – and this is an important limitation – this is happening as long as companies continue to build systems in the way they currently do. I should note that there are now many options built into our devices that can reduce interruptions. It is, for instance, possible on many devices to configure a silent mode to have applications that do not send notifications and do not display missed messages. Device *capabilities* do not, however, necessarily translate to *usage*, as historically, 95% of users have been keeping manufacturers' default settings.[141]

4.5 Digital Juice

The machine life of our online services and devices that I have described so far is driven by backend systems: the warning signals on our cell phones, the shimmering vibrating surfaces, the likes and retweets I see come from other platform users or the billions of virtual bots populating the Internet these days.[142] Advertising and content is funneled to us via marketing agents working for whoever pays their bills, whether that be political campaigners or shampoo advertisers. There's an invisible world of others out there in the Internet's underbelly that accesses us users at all times.

But there is another digital quality that binds us deeply beyond that shadow economy out there: I want to call this quality the "juice" of the digital.[143] This juice is referred to in professional jargon as "amplified positive feedback." When we interact with our services, agents, etc. then they typically respond by means of color play, movement, sound and speech in an extremely positive way. In other words, the digital reacts to us with delight. For example, when we walk across bright green grass in a well-designed virtual world, we can see this virtual grass bending under our virtual feet, the way that the shades of color change, and the sound of it rustling. These effects are much stronger than we could ever experience in the real world – in this sense they are "amplified." Another example is the "trash can" feature on Apple devices. When I delete something and throw it into the virtual wastebasket, I hear my documents literally dropping to the bottom of the wastebasket; a sound of rustling and the slamming of a lid signal the deletion process. In short, I do something very simple and get an overly strong positive feedback to my action. Through such simple mechanisms, I constantly get from the digital world what the food industry calls "flavor enhancers." And that feels awesome – just like juice.[144]

4.5.1 Triggering Power and Achievement Motivation in Users

The fact that we humans all react to this digital juice in a way similar to how we react to alcohol is again due to our psychology. In motivational research from the 1980s, David McClelland described three important motives for humans that are present in all cultures of the world: power, achievement and group affiliation.[145] The power motive is about the more or less pronounced desire to be able to control the world and give it structure, to be able to move things as one considers appropriate. Power-motivated people are often competitive: they want to win. Achievement motivation is different. Many know the feeling of wanting to bring something to its conclusion: for example, pushing through the pain of the last few kilometers of a cross-country skiing race, even though your legs are already hurting. Achievement-oriented people go to bed satisfied in the evening when they feel that they have done something and diligently completed the day. And finally, there is the very strong motive for many people to have a connection to a group, to feel cared for and to be appreciated. This is the affiliation motive.

Digital services have the uncanny ability to correspond to all of these three human motives. First, they give us users a sense of power. When we go online and change things to the way we want them, we get the impression that we have power over things and that we have done something. For example, casting a wand in an online game gives us a sense of power that we just don't have in the real world. If we can say that Amazon Alexa should "play music" and Alexa plays music then this gives us the feeling that we are powerful. We possess a digital power even beyond real things, something we used to call magic in earlier times. Equally, if we throw something into our digital wastebasket and a rattling noise gives us the feedback that everything has been well completed then this appeals to our achievement motive, which wants to bring things to a close.

One of the biggest temptations the digital has given us through its overly positive feedback is the narcissistic mirror. In the real world, you have little power to always look good or to win: you have red eyes from hay fever, the pimples on your face erupt unexpectedly, your voice fails you at a public appearance, and when you tell a joke, it doesn't land how you wanted it to. You are now freed from all this misery in the digital world, where you can always look good and win (most of the time). Of course, you post only the 20th photo of yourself in which you finally caught the right angle. The pimple will be retouched with AI. The joke will be reworked with AI as well before putting it online. And then when you look with pride at your own narcissistic mirror, the Instagram or TikTok wall of your own success and beauty history, you naturally feel great. It's like an acknowledgement that you indeed possess all the values you have ever aspired to. You wanted to look beautiful: voilà, you can see in the digital mirror that you are beautiful. You wanted friendship: voilà, your photos are populated with friends. You wanted success: here's the photo of that last business trip to New York. You wanted fame and honor: here's the post of you onstage at that last conference. The digital nourishes the illusion that we have power over the values to which we aspire. It makes us believe in achievements we never actually lived up to.

The result is that many people are building a separate and artificially enhanced "e-personality" online.[146] As these e-personalities groom their online looks and stare like Narcissus into their self-created online mirror images, they start to believe that the whole great network "out there" sees them and appreciates them for all the great things they seemingly have to offer. Seemingly, their longing is even fulfilled, because when the echoes of the virtual social world reverberate back to them through likes and hearts, retweets and comments, they feel confirmed in this beautiful e-personality. The positive echoes of the virtual world reinforce their belief in their self-made e-personalities. And since these are so easy to maintain, they begin to believe that life in general is as easy as online. Thus, in the real world more and more morbid effects of self-exaggeration and overestimation of people's power and performance arise. Expressed in terms of values, there is false pride, hubris and arrogance, as well as a lack of humility, respect and modesty.

Doctors are particularly confronted with this problem. Stanford University psychologist Elias Aboujaoude, who has studied e-personalities for over a decade, speaks of a pathological "grandeur." He describes how people who have studied their health symptoms online come into his practice and are deeply convinced that they are more knowledgeable than he – a man who has seen thousands of patients. In social psychology, this phenomenon of false grandeur, the Dunning-Kruger-Effect, has been christened "meta-ignorance." It has two consequences:

first, individuals believe that they are better informed than others; and, second, they are unaware of their own misperception in this regard.[147]

Incidentally, the phenomenon of meta-ignorance does not only exist in the medical field. Even when it comes to computer science, today there are many self-proclaimed "experts" who want to help shape our technical future and are actively discussing AI. They dare to participate in the highest political advisory boards, but never before have they seen a system from within, nor ever modeled a system, let alone programmed one. Therefore, they know nothing about data flows, communication protocols, technical architectures, information qualities, requirement analyses or development processes. They know neither the technical boundaries and development paths nor the work of development teams. They also have no knowledge of statistics, algorithms, data and their limitations, and therefore unfortunately do not notice how much nonsense they talk about when they talk about AI.

4.5.2 The Illusion of Affiliation and Community Online

Big social media platforms focus not only on the satisfaction of power and performance motives. There is another human vulnerability they have learned to exploit: our desire for belonging. Perfectly staged e-personalities on Instagram, TikTok or YouTube videos and blogs give us the feeling of gaining acceptance and friendship. And if there are many friends and followers who signal this to us with their likes and retweets, then we sometimes feel more comfortable with our digital selves than with our real self.

Noteworthy here is the role of others, whose presence exists more in our imagination than in any reality. As I have already described, research shows that human intelligence is built in such a way that we not only perceive what we see directly but that we always autocomplete the world around us with our imagination.[148] So if we're on a social media platform, probably only 5% or less of our friends or followers will be there with us at the same time. However, we are implicitly assuming that the other 150 followers on Instagram will somehow see what we post.[149] We imagine the fictional co-presence of our friends or followers as real, even if the statistics prove otherwise.

Being close to others is hugely important to us. Neuroscience research has proven that we humans experience social exclusion in the same pain center in the brain as physical pain is felt.[150] Loneliness, it therefore follows, can physically hurt. We strive to avoid this pain. And so, it is hardly surprising that we feel drawn to a social network like bees to a flower. "People really crave social affirmation," says Erin Vogel, a psychiatrist at the University of California, San

Francisco.[151] In some European countries up to 50% of people live alone. As a result, the lure of searching for company online is very hard to resist.[152] However, the sad truth is that most "friends" do not even look at the posts of their acquaintances. Only 7–17% of our time spent on social media is invested in viewing the posts of friends.[153] Instead we prefer following influencers or hanging out with our favorite AI avatar. So the "community" is, to a large extent, nothing but imagined. Long-term studies of social media hence come to the conclusion that it is only beneficial for people who have strong ties in the real world. Those who need the digital medium to connect with others become unhappy and are more likely to fall into depression.[154]

An open question as of yet is what will happen to us as we get attracted to artificial chatbots and personal agents, even more than is the case already. As of 2025 there are conversation limits set by some services such as ChatGPT.[155] The limit is set up for cost reasons, but once competition forces AI providers into unlimited conversation, the question is how real human bonding will evolve. There are multiple reports of people already getting married to chatbots.[156] Others get romantic tattoos to show their affection for their virtual assistants.[157] In one case, a chatbot successfully encouraged Jaswant Singh Chail to break into Windsor Castle and attempt to kill Queen Elizabeth II.[158] The first known suicide encouraged by a chatbot was of a Belgian man in March 2023.[159] Other examples followed, including Sewell Setzer (14 years old) and Adam Raine (16 years old).[160] While these interactions ostensibly only constitute a small proportion of interactions, the popularity of generative AI means that a huge number of people are potentially bonding and trusting to a questionable extent. They are getting entangled with illusions.

4.6 Being Entangled in the Web

Numerous interviews, group work and surveys undertaken as part of my courses at the Vienna University of Economics and Business have enabled me to gain a deep insight into how the described mechanisms are driving digital natives into digital dependency and how their thinking is changing. It's amazing how fast all of this has come about. Actually, we, the builders of the Internet, had hoped that people would move around it like spiders, reweaving the social fabric and developing new forms of democracy, participation, openness and global freedom. Instead of spiders who know how to weave the Web, many have become flies that are trapped inside it and seem to dissolve. Flies cannot help becoming more and more wrapped up in the Web. After a short while they are so intertwined that they cannot free themselves from it without help – just like the student who told us in an interview study that the loss of her cell phone would feel like she had her hand chopped off.[161] Such a statement is far from an isolated case. In an American study, two-thirds of the participants said that they would be ready to have their cell phones implanted under their skin.[162]

Normally, an intimate connection with technology does not have to be threatening. As a wearer of spectacles, I know that taking off my glasses feels like missing a part of my body. Technology as an extension of the self has often meant progress for humanity. There is historical evidence that we would never have developed as a species into the present form of human intelligence had we not for the past 2 million years used technology. This is how our brains actually developed.[163] However, so far, we have always been free to choose whether we wish to use such technological aids. We were able to determine the moment and duration of access. If ever I felt that I'd had enough, I could simply switch off and do something else. Equally, in the first phases of digitalization, for example, we simply switched off the computer. This has changed. Today, our integration into the digital network is not so freely selectable. Instead, the Internet proactively envelops us; it calls us, beeps at us, blinks at us and keeps sending us reminders. Devices are always on and on standby, reachable by default in real time. One subject told us in an interview: "What I'm noticing with myself and with my friends is that we cannot control it, that is the problem."[164]

This uncontrolled access behavior occurs because many users are pushed into a passive participation mode. We allow the Web to keep us busy. The British network specialist Julia Hobsbawm writes, "If we were once users of networks, travelers in trains, boats and cars, people who have used phones, then now our lives are embedded in networks and we behave as if we were actually a part of them."[165] In my students, I observe this level of entanglement with the Web. One told us, "Even if you're not there, you know everything, you feel like you're there, because, you know, you're kind of like everywhere, you can be everywhere."[166] My student and Julia Hobsbawn describe the same phenomenon, outlining what it means to become part of the Web. People no longer feel like locally present entities deciding their own life priorities, but like entities that merge with the Net, triggering a kind of fugue state of constant passive ubiquitous responsiveness.

Many fans of digitalization would probably argue that this merging of humans with the Web is nothing to be afraid of. The benefits of networking are so great that they lift humanity up to a new level of evolution.[167] Perhaps we should see it as progress to not be so locally bound any more; to be with everyone, everywhere, all at once. But I wonder what that does to our personalities? What effects does permanent unpredictable access to our person have? At this point I would like to share the following story from the satirical magazine, *The Onion*:[168]

> GAITHERSBURG, MD – Local man Marshall Platt, 34, came tantalizingly close to kicking back and having a good time while attending a friend's barbeque last night before remembering each and every one of his professional and personal obligations, backyard sources confirmed. While cracking open his second beer as he chatted with friends over a relaxed outdoor meal, Platt was reportedly seconds away from letting go and enjoying himself when he was suddenly crushed by the full weight of work e-mails that still needed to be dealt with, looming deadlines for projects that would take a great deal of time and energy to complete, an upcoming wedding he had yet to buy airfare for because of an unresolved issue with his Southwest Rapid Rewards account, and phone calls that needed to be returned.
>
> "It's great to see you guys," said the man who had been teetering on the brink of actually having fun and was now mentally preparing for a presentation that he had to give on Friday and compiling a list of bills that needed to be paid before the 7th. "This is awesome." "Anyone want another beer?" continued Platt as he reminded himself to pick up his Zetonna prescription. "Think I'm gonna grab one." Platt, who reportedly sunk into a distracted haze after coming to the razor's edge of experiencing genuine joy, fully intended to go through the motions of talking with friends and appearing to have a good time, all while he mentally shopped for a birthday present for his mother, wracked his brain to remember if he had turned in the itemized reimbursement form from his New York trip to HR on time, and made a silent note to call his bank about a mysterious recurring $19 monthly fee that he had recently discovered on his credit card statement.

I think a lot of people who are living an active life today and have to be online a lot due to their job can profoundly relate to Marshall Platt's situation. It is a state of being nowhere, to not breathe the oxygen of real presence and to constantly be forced to think of other things than those before you. To hear an inner voice that is constantly talking. In Buddhism, this phenomenon is called "monkey mind."

The Web has woven us into a vast connection with strangers, institutions, documents, events and now even avatars. And these other "nodes" are accessing us at all times and from all sides, absorbing our attention and our time. But it goes further: no sooner have we satisfied one node in the Net, another one appears to pull us in. Since we do not always know exactly what and by whom the next request will come, the entropy of our lives increases massively. Entropy is not neutral for us. It is a measure of the chaos and disorder in living systems. My own experiments at the Vienna University of Economics have shown that entropy brought about by the consumption of digital media has a negative impact on our human creativity and personality. In a study of 134 participants, I was able to show that increased entropy on a computer screen initially leads to more activity, more words, more output (verbosity).[169] However, upon closer examination, the ideas or statements in this output were mostly meaningless. What's more, the relation to oneself – the use the of words "I" or "me" – were significantly reduced with increased screen entropy. It seems to me on the basis of my own experiments that people are losing their capacity for self-reference when entropy increases.[170] These observations are troubling indeed. In particular, I see fewer and fewer people use the words "I" or "me" in e-mail and messages. Instead, they write "have no time" or "am there", as if the sense of self had evaporated. Such developments are not banal. Language is of the utmost importance, and is the key to understanding what is going on inside us. As philosopher Ernst Cassirer once wrote, "We know this 'life' only in its 'expressions'; but precisely this is the quintessence of our whole preceding observation that 'expression' is nothing accidental, inessential, 'external,' but rather that it is the necessary, the true, and the sole manifestation of the 'inner' and of the essence itself."[171] If we no longer use the word "I," does that mean that we lose ourselves in the digital realm?

4.6.1 The Development of Shallow Personalities

Personally, I struggle daily to slow down the demands that come from the Web. I try to give myself the space to work on everything in peace and quiet. Nobody can tell me that I have to answer in real time! But even after completing my usual ritual of turning off my devices, setting all presets to low, and even meditating to maintain my inner balance, I have the impression that very few of us, myself included, are able to free ourselves from the power of the Web. Friendly "reminders" reach us permanently, from everywhere. They are like soft electric shocks that ask us to work here and now, everywhere, at all times for anyone who sends us something. As deeply social beings, we feel obliged to be open and receptive to this madness. But this openness may have profound implications for our human identity, which dissolves in a state of permanent passive responsiveness; not only figuratively, but physically. What happens here? To explain, I want to retrace a few arguments.

Research has shown that unlike animals, we humans build an autobiographical identity throughout our lives. We do this by not only paying attention to things in our environment but also assigning their meaning.[172] The meaning of things and events memorized is unique for each person, and I would like to hypothesize that it has a lot to do with the values that we have learned to attach to things and events. My coffee cup, for example, has a very different meaning for me than for my husband. Although we both drink from this cup, the inner meaning I associate with it is unique. It is associated with the value of friendship, because I remember sitting with my friend on a balmy summer's evening in an inn where this mug was on sale. She gave it to me as a symbol of our friendship and told me that it should inspire me when writing my book. For me, this cup has become my daily work companion: a symbol not only of that friendship but also a symbol of inspiration and a concrete reminder of the joy I felt on that evening. The meaning of the cup goes far beyond its practical purpose of drinking. Why am I describing this in such detail?[173]

The meaning we give to things forms part of our personality. The places we visit, the people we meet and admire or hate – all become woven into the tapestry of our selves. Each of us, thereby, develops a unique and meaningful history that is stored in our long-term memory, in our cognitive schema. Places and people become the repositories of valuable memories, just as my cup overflows with memories of friendship. And as a result of experiencing so many valuable things in the world, we are then able to attribute to each present and new thing that comes to us an individually unique meaning: our memory mingles with the newly perceived and enriches our own singularitas. In contrast to lower animal species, it is clear to us "what memory is there for." We are human beings because we can use our valuable memory in the service of thinking – a process that allows us to give meaning to the world around us.

However, the prerequisite for building this memory apparatus and its ability to generate meaning is that we actually pay attention to the things around us. If I had not listened to my friend when she handed me the cup, and instead turned to an important mobile message, then I would not now have my morning coffee full of friendship. The mug would be reduced to a meaningless vessel. The situation would not only have been ruined at that time but I would also in the here and now lack a valuable node in my memory.

So, if the Web is constantly showering us with virtual impulses from everywhere and nowhere, then we probably cannot cultivate our autobiographical memory as we used to.[174] The meanings that enrich our consciousness turning things into value-bearers become more superficial. Our personality, as Nicholas Carr so pointedly described, becomes "shallow."[175]

4.6.2 The Loss of Life Energy

Shallow personalities may still be endured. However, as psychologist Arthur Glenberg describes in his fundamental article "What Memory Is For," a second phenomenon is also at play here: There are two ways in which people access their memory. The first is purely reactive and automated. In such cases we do not have to think about the process that activates our memory. For example, after a certain number of driving hours with manual gears everyone is able to automatically use them without thinking.[176] These kinds of automated routines are also present in animals. In the second form of memory access however, which distinguishes us from animals, we think actively, rather than following a routine. We know these moments, in which we pause to consider what to do next. For this reason, driving a manual car is initially difficult. We need to think about the correct use of the gears, the pedals, etc. It requires our full attention. In such moments, we need to fully focus on the situation and try to block out anything that is not relevant to us for solving the challenge. This degree of concentration and blocking out of other "stuff" is extremely exhausting. And this is precisely where the problem arises: these efforts are becoming more and more difficult for us the more we live in a digitally networked fashion. Due to the permanent reaction mode and the digital interruption mode, our ability to block out our environment and concentrate fully on the matter at hand is decreasing. This means that a deeply human form of thinking becomes increasingly difficult: that of concentrated reflection. A recent MIT study found, for instance, that using a chatbot for assistance significantly decreased the intensity of brain activity and hence connectivity. The loss of brain connectivity again inhibits cognitive development and ability.[177]

What will we humans become, then, if our memory shrinks, and the richness of meaning of our physical environment fades away. And how will we function if we have less and less power to think because we are permanently occupied with "other voices from other rooms?" We become, according to Stanford professor Robert Hanson, "borgs."[178] Borgs, a race featured prominently in the Star Trek series, are entities made up of a synthesis of synthetic and biological material (see Fig. 4.1). All Borgs belong to a large community; they are no longer individuals but part of a civilization that exists as a great collective consciousness. Every Borg hears the voices of every other Borg. Nobody has his or her own voice. It is worth taking a look at the Wikipedia description of the Borg culture because it is eerily similar to the transhumanist discourse so prevalent in our society, which I will describe in the next chapter.[179] The Borg Society is the model of a totalitarian but nonhierarchical society. Their concept is the permanent technical adaptation of their bodies to changing environmental conditions and expansion throughout the universe at all costs. The Borgs continue to evolve by "assimilating" other races and their technologies; that is, absorbing their knowledge and experience as a whole in order to agglomerate to themselves, through a collective consciousness, new qualities from that community. Individuals and individual consciousness do not exist among the Borgs. Assimilated individuals are turned into drones, whose bodies are equipped with mechanical implants and nanotechnology. Being thus connected is a paradisiacal state for the Borg because the collective consciousness ("hive mind") of the species has become virtually immortal. Nobody from the collective is left alone, everyone is needed. Perfection is the ideal of the Borg collective. The symbiosis of technical and biological components represents a step towards this perfection.

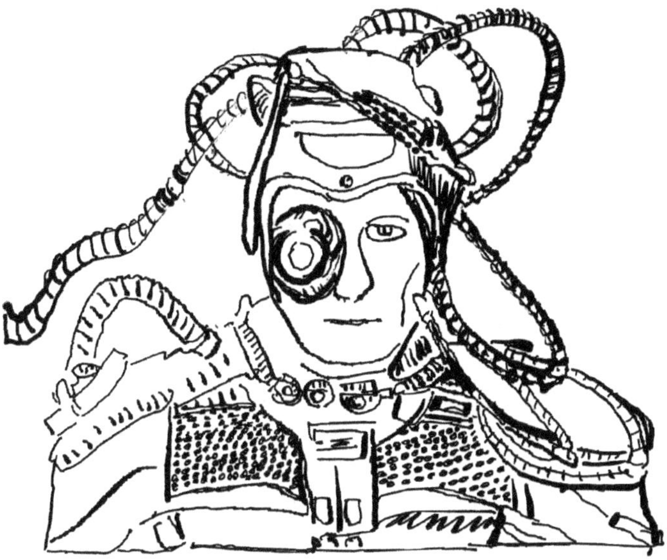

Fig. 4.1: The Borgs as a new type of human (own depiction).

4.7 Value Consequences of our Digital Transformation

I have described the nature of the digital fabric as well as the Web's business dynamics. I have laid out how the digital fabric is doomed to incompleteness when compared to the singularitas of human reality: it creates a division of content and form; it is, at least for the foreseeable future, still prone to error. For no reason we have incorporated into our daily lives a complex entity that robs us of our human vitality. Though the clock is ticking, we still have the freedom to work for human progress in the digital age. But it is high time that we become more aware of the consequences of digitalization for human values, which I will now attempt to summarize; specifically the value trade-offs we need to consider.

4.7.1 Efficiency Versus Dignity

That the digital lacks singularitas is, in some sense, also an advantage. Human contacts and communication processes that are digitally mediated are often associated with less friction. They are simpler, more direct, and simply faster. Real human encounters, in contrast, can be time-consuming and exhausting. Digitally we can avoid this. In particular, shy or insecure people benefit if they can linger behind the barrier of digitally mediated communication and do not need to fully expose themselves. People with busy careers may also save a lot of time and gain new freedoms. These are all positive value qualities that we should not underestimate and that contribute positively to our appreciation of the digital medium. The progress curve (see Fig. 2.2) is trending upwards due to these benefits.

However, one should not underestimate the negative values of digital incompleteness and insubstantiality appearing at the same time. Massive adoption of digital processes and virtual communication in companies is highly problematic, precisely because it is the richness of interpersonal relationships that are at the heart of corporate success. This is true above all for HR departments, for many purchase functions, for customer relationship management, for sales and all service areas. In our various roles as employees, customers and also as individuals, we wish to be seen in our singularitas. This is central to our sense of purpose and dignity and hence to our work motivation. As I showed with the dating example, platforms that reduce people to homogeneous entities diminish their dignity and uniqueness. They exchange these two high values of dignity and uniqueness for efficiency.

An undermining of dignity can also be caused simply by the described incompleteness of digital communication in e-mails and messaging. In Chapter 8, which deals with freedom in the digital age, I will describe in more detail how splendid rituals of respect served our freedom in the analog world. Our relationships were marked by courtesy. Today, digital communication offers unprecedented efficiency, flexibility and leanness in direct exchange; perhaps it is also more environmentally friendly (because we use less paper) and allows us to stay in contact with others more easily. Not infrequently, however, this form of easy communication also creates misunderstandings, confusion and conflict, resulting in unnecessary, energy-consuming shadowboxing and undermining of community. These negative values – conflict, loss of confidence, rudeness and disrespect – are the downsides of the incompleteness of the digital. The progress curve goes down steeply if we are not aware of these value costs.

Is it possible to still achieve a positive overall balance? Are we able to take advantage of the indispensable benefits of the digital fabric while deliberately stopping the negative effects? Everyone can start on a personal level by seeking as much direct and analog conversation as possible, and using digital communication only where it is advantageous for temporal, spatial or factual reasons, for example, when making an appointment with one's hairdresser or when cooperating on a project remotely.

But when it comes to managers, those in senior positions should think about this advice and not hide behind the efficiency of their e-mail, instead embracing interpersonal communication with employees. Businesses should consider wisely in which areas digitalization strategies could be counterproductive, eroding human singularitas. Wherever singularitas plays a role, automation and simplification should be questioned.

4.7.2 Trust Versus Caution

Mechanical machines make fewer mistakes than we humans do in many areas, and we've come to love the seamless ease of their use for a wide range of purposes. What is more, unlike humans, where fatigue slowly decreases the quality of work, machines are never tired and can continue to work flawlessly for as long as is necessary. So, there are definitely positive values in play here, such as precision, reliability and performance, values fundamental to production and engineering. However, with the advent of digitalization and the use of software that is often opaque, mediocre, ill-documented and sloppily-developed, relentless "error-gremlins" have crept into all corporate areas. These error-gremlins are something we never expected and dismiss as folklore because in the history of technology over the past 400 years we have learned that machines are largely error-free until they finally break down.

To avoid being compromised by these errors we have to take action on many levels. It's not enough to just get software developers to debug: they need to develop much more transparent systems that are better documented, modular, and allow for more control over all built-in objects, scripts, databases, AI components and data sources. Anyone integrating a component should actually have some kind of access to what is being sourced. Here, of course, uniform technical access standards are missing or are not developed with enough detail.[180] Another main impediment in the corporate world is a lack of developer time. For investors it is counterproductive to put development teams on a schedule that is too tight and does not allow them to devote sufficient attention and also autonomy to the quality of their systems.[181] It is not enough to give time for a system to run reasonably well; time must also be allowed to build in control and transparency. Most importantly, outsourcing of software development, the use of unqualified developers that only are good at AI prompting (instead of programming), the massive integration of external untransparent software components, etc. are all developments within software development that undermine the trustworthiness of software products. Political forces could help here by demanding systematic certification of software systems across the board. When it comes to error liability, IT providers must of course be made liable for all defects and the path to court must be easy for victims, always assuming first that the software made the mistake and not the human. Software providers should be forced to publish information on the accuracy of their machine output before launch; the respective metrics could be displayed in the form of a virtual "packaging insert," just like nutritional information is printed on food packaging.

That said, no matter what measures will or will not be taken, governments should be careful not to expose their citizens to the digital error-gremlin, especially when it comes to e-government systems, border and security systems, and public services. I would like to highlight this point because these e-government systems can cause much bigger ripples than any small app that may make a few mistakes. Nearly 100 million citizens live in Europe's German-speaking countries. If a mistake affects just 0.5% of those people, then we're dealing with 500,000 digital victims suffering from IT errors. This is more than hypothetical: the "Robodebt" scandal in Australia saw false debt notices issued to approximately 526,000 people as a consequence of an adjustment of the calculation method. On top of the alleged overpayments, a 10% "recovery fee" was even added. Ultimately, rather than saving money through an automated system, it ended up costing the government AU$970 million.[182] Such developments bring about destruction of value and a growing loss of confidence in governments. Similarly, the stepping down from office of the whole ruling cabinet in the Netherlands after the child-benefit scandal (see above) should be a warning sign for governments around the world.

That said, we again have a chance to deal with the error-nature of the digital fabric and turn the progress curve upwards. In addition to the measures described above, it is crucial that we better understand software's vulnerability to error and communicate more openly about it, particularly when it comes to the use of statistical methods and thus of natural error probabilities in our calculations. We have to be much more realistic. Every teenager should learn at school how machines are susceptible to errors. Everyone should understand some statistics and hear about error probabilities. If everyone is aware of these natural limits of software and AI systems, then everyone is simply more realistic in how far they rely on them.

Finally, in sensitive human areas where we do not wish to forgo the digital, for example, in medical diagnosis, computers should always publish the error probabilities of their results. This would foster a scenario where physicians are more realistic in their assessments and do not unquestioningly accept the judgments of the machines or AI systems. In fact, practitioners should learn about typical machine mistakes. They could be involved in tweaks and improvements to the machines. Healthy competition between manufacturers could arise: who can produce the devices with the fewest bugs and lowest error probabilities? In this way, we could perhaps even turn the error-gremlin into a motor for better innovation.

4.7.3 Power Versus Addiction

An important question is to what extent the "machine life" of our technical devices really benefits us. Amplified positive feedback makes us happy because we like the small dopamine doses of superficial happiness that the digital brings to us when it offers a nice "user experience." We also really like anthropomorphic AIs. On the other hand, I do not see any positive societal values that should result from this specific inherent life of our digital devices. When devices proactively turn to us, or when they proactively share data in the background (without our knowledge and consent), it is really only the service providers who benefit. This is baked into today's Internet economy, which is built on trading and exploiting our personal data and attention. This should not be the case.

Digital service providers are gaining outrageous power and control over all users who are hooked on their services, allowing them to develop ever more powerful systems, such as AI chatbots. This is of course very tempting. But the social cost of these business models and the addictions they cause are also outrageously high. I would argue that many users experience addictive behavior, loss of time, loss of cognitive power, inability to concentrate and passivity, as well as a loss of the flow that makes people happy, and encourages creativity, empathy and productivity.

Against this background, governmental intervention is long overdue. Recently, governments have at least started to ban social media and smartphone use for kids. This is a start. But in my view, the banning of smartphones for kids and the abolition of addiction-driving business models is at least as important. In truth, it is these business models that lead to the consumption of our attention, our vitality and our personal information. When services are designed to be intrinsically addictive (such as irregular feedback or endless loops) they should be prohibited by law, and the factory settings of our devices should be as privacy-oriented as possible and adjustable in a fine-granular way by users. AI-based personalization should not be available by default. Users should be able to switch off any data- or AI use without being excluded from service provision. That said, service providers might then also need to charge for their services and they shouldn't be blamed for it. You get what you pay for.

At the same time, it is paramount that all educational institutions and companies offer media training that problematizes digital addiction and that consequently helps people to train their willpower. At Vienna University of Economics and Business I suggested that smartphones should be banned from campus and communication zones should be set up, just as we have smoking and non-smoking areas today. Incidentally, it is noteworthy that tech elites are already going in this very direction when it comes to their own children. They have been protecting their children for a long time from the digital service world. Steve Jobs banned his children from using the iPad he had just launched on the market.[183] Bill Gates did not allow his daughter to receive calls on a smartphone until the age of 14.[184] In Silicon Valley, many elites send their children to schools where low-tech chalk boards are a competitive advantage. In principle, it is perverse to see how tech elites educate their children to abstain from the very technologies they push on billions of people every day; indeed, the rather piquant suggestion has recently been made that tech elites should be forced to have their own children be the first guinea pigs of the services they offer to the public. That said, as of 2025, the first government officials are beginning to wake up. In Texas and quite a few other states in the US lawmakers prohibit students from using personal devices on district premises during school hours. And this not only includes smartphones but also tablets, gaming devices, smart watches, Bluetooth earbuds, fitness trackers with messaging, personal laptops, two-way radios, pagers, and more. So while medical needs and special education accommodations do allow for the benefits of digital services, social interaction as well as identity formation can flourish again through analog encounters.[185] Some European states are thinking about similar measures.

4.7.4 Availability Versus Freedom

Hearing Mark Zuckerberg speak euphorically of his vision to "connect" everyone, I have always wondered what *values* he actually wants to foster by doing so. I have already made clear that networking is not a value in itself. It is just a technological capability. I suspect that Zuckerberg's idea is based on the assumption that progressive networking could lead to more peace and friendship? An interesting theory that could be explored in a research project.

An immediate and likely positive value is, of course, the increasing accessibility of people or reachability; perhaps findability. At the beginning of this book I already described how the Internet in its early days suddenly allowed me to get in direct contact with scientists at MIT. Even if this is not the case any more, digital networks save a lot of time, physical strength and, of course, money in communication. Easy accessibility is extremely convenient for many who work from home or abroad today and would otherwise need to move. All these benefits are, in my view, a true gain that I personally would not want to miss out on.

Beyond these value gains, however, my analysis shows that it is precisely this accessibility, with its direct, unbuffered access to each and every person, that breeds negative values as well. The negative values associated with networking include the loss of freedom and control over one's own time. In addition, a loss of attention, loss of presence, superficiality, passivity and, in extreme cases, even a loss of personality. Technically, I cannot imagine how we might be able to preserve the benefits of networking while curbing the negatives. One possibility would be to slow down access to the Internet. For example, we could consider the artificial incorporation of latencies that take speed out of the Internet. For example, governments could prescribe that e-mails are delivered only once a day; perhaps not at all on Sundays. Social networks could update their postings only once a day. Software systems could integrate feedback messages along the lines of "out-of-office" replies. Some of my students have even suggested that such measures should be required by law in order to achieve social synchronicity, so that everyone is offline at the same time. Of course we all need to think about such ideas more intensively since the advantage of accessibility should not be lost and many online users would perceive such guidelines as a serious infringement of their freedom. I therefore believe that education and awareness of what digital networking means to us individually is paramount. It must be clear how much time and energy it costs us to be digital. It must be clear to what extent it restricts true community, distracts, limits the framework for one's own development and learning and, through real-time access to our person, gives rise to feelings of loss of freedom. All these are aspects that should be considered – and for which it will take courage and vision to develop action strategies.

5 Our Modern Understanding of Progress

The analysis of the nature of the digital has shown that we need to be careful with digitalization. So far, many positive values have been unleashed in our society: new time-freedoms, increased flexibility, mobility, accessibility, professionalism, access to knowledge and enjoyment of services such as photo and music sharing, social networks, video calls and AI support. However, it seems that these benefits are eaten up by a series of values that are undermined. Effectively, a cost-benefit analysis of digitalization might be needed. Value costs include the emergence of addiction, loss of time, loss of power, inability to concentrate, passivity, flow losses, conflicts, loss of confidence, rudeness, disrespect, loss of dignity, loss of attention, loss of presence, loss of empathy, superficiality, changes to the brain and eventually even the progressive loss of personality.

So, how should we continue with digitalization at such a pivotal point in the history of progress? Future visions of the IT industry revolve around constantly evolving technologies since it is taken for granted that, of course, everything imaginable will be automated and everything human is, at least as far as is possible, replaced by the artificial. But why? Here, I will circle back to my initial question: the *why* question. Maybe car mechanics *like* to assemble cars in the production lines of automobile manufacturers? Maybe people choose front of house because they *like* working on reception? Employees in a human resources department may be *interested* in interviewing job candidates. And without a doubt, many taxi drivers *like* to drive a car! So, why do policymakers see the automation of human activity as intrinsically progressive? Isn't human life really about activity?

I must confess that a strange feeling strikes me when I imagine that in the future, FoodIS couriers might receive their delivery orders by phone again, like taxi drivers used to in the past. The computer voice of the AI seems, to my intuition, much more advanced. I can well understand that the Austrian farmer, Johannes Gutmann, was considered a madcap when he started to work the soil manually to grow the Sonnentor herbs. I expect that many readers of this book will dismiss my suggestion to re-manualize some parts of industries as naive, because it is simply retrograde. But is that wise?

Why do we always perceive technical upgrades to be automatically better than traditional ones? Where does this immediate conclusion come from that new or digital = good, and in contrast, existing or analog = bad? Most importantly, why do we intuitively insist on this mindset when mature existing processes in companies or other institutions (e.g., schools, government agencies and hospitals), as analog and manual as they may be, often lead to better qualities and healthy returns?

 | https://doi.org/10.1515/9783111677736-005

More realism is required. Anyone who plans a larger IT investment for their organization should know that only a fraction of innovations are really so successful that companies retroactively rate them a success. Estimates in fact vary between 25% and 61%, depending on the project.[186] Also, the fear that one could fall behind one's international competition without digital upgrade seems questionable. In truth, true disruptive innovations changing industries are rare.[187] If you look at app usage underlying many start-up businesses, a sober analysis shows that of the 4.83 million apps available on Apple and Android, only 10 are used on a daily basis by consumers, and 30 monthly.[188] Added to this is the issue of responsibility for one's own company, employees and society: if entrepreneurs know that up to 47% of all jobs could fall victim to automation, then they should also see that this is not just a political problem but also one for their own company, because it is inevitably linked to a loss of purchase power.[189] Against this backdrop, the firing of call center staff and cashiers starts to look a bit dubious. Yet, I do not want to discuss these open questions that other authors have expanded on really well. My concern is only this: if there are serious arguments for questioning technical upgrading and digitalization, why do we do it? Why is it that almost all members of today's elite are caught up in a digitalization whirlwind, with the simple mantra of digital transformation = progress?

I do not think our automatic reflex that digital transformation = progress is solely due to the storytelling of the IT marketing machinery (which I described in chapters 1 and 2). Instead, we are following the well-trodden ways of our forefathers on a path of progress that they chose a long time ago. This is a path that we no longer question. It understands "progress" as technology-centered, not human-centered. A technology bias is firmly anchored in our minds that prevents us from questioning and reflecting soberly on the new technology-breed that has now emerged with the digital world. This prevents us from thinking about how real human value creation can be achieved, in particular when it comes to AI. As a result, we are losing qualities that we actually should preserve.

For me, the relationship that many people today have to technical advances looks a bit like a longstanding love affair that knows a serious crisis is looming but refuses to accept it. When such relationship problems arise, then we can either face them or continue as before. When faced with the challenge of trying to rescue the relationship, a good proactive response would be to go to therapy and undergo a process that elaborates deeper beliefs and problematic behaviors; behaviors that often stem from our own history. In this chapter, I would like to address just such a therapeutic solution for our checkered relationship with technical progress.

5.1 Modern Beliefs in the Nature of Progress

The origins of our understanding of progress today are rooted in the High Middle Ages. Here, the concept of "inventing" and "progressing" appeared for the first time in the *Metaphysica* of the monk Albertus Magnus, also known as Albert the Great (1200–1280). He wrote: ". . . *ut ab inventis ab eis ulterius proficeremus in philosophia.*"[190] Translated, he said something I would underscore when thinking about positive progress today: ". . . that we strive in pursuit of the wisdom of what our forefathers have uncovered and invented."[191]

It is important to know that Albertus Magnus had a classical understanding of education, which dominated for over 1,500 years. Progress from this perspective meant that *each individual* progresses in the course of his or her life by developing as a person. Foucault has described this personal progression also as "aesthetics of existence." It is about self-actualizing through a cultivated life (see Fig. 5.1). Foucault interpreted cultivation as a concern for oneself – concern for one's own health, sexuality and the gaining of knowledge.[192] In the Aristotelian tradition that Foucault discusses, it was important for everyone to become aware of their particular talents and purpose in life: their own *telos*.[193] These talents indicate a specific role one can fulfill within a community. A talented speaker will be able to entertain a group. A gifted leader will be able to guide them. A loving person will be able to look after them. A *progressive* person, meanwhile, is one who correctly recognizes his or her talents and tries to find his or her natural place in the community. That is why the entry portal to the oracle of Delphi contained the inscription and exhortation "Know thyself!" A progressive life could only succeed if we "philosophically" – in other words: reflectively, contemplatively – recognize our own place in the hierarchy of nature, the cosmos, time, things and living beings around us and behave accordingly.[194] Of course, respect for the traditional order of things is important in this endeavor. The monasteries at the time of Albertus Magnus preserved and studied them in the *artes liberales* and contemplated them in meditation. It was considered to be really advanced or even wise to have one's own will and strong abilities harmonize with this order of the universe.

5.1.1 Progress Means New = Good

In the 13th century, when Albertus Magnus spoke of "inventing," it was hence based on a philosophical tradition whereby progression (*proficere*) was synonymous with the disciplined striving (*philia*) of each individual for wisdom (*sophia*).

From this perspective, the new should not "disrupt," but harmoniously embed itself in the natural order of things.[195]

It would have been nice if this thinking about progress had lasted, but these ideas would not be dominant for much longer. Already in the time of Albertus Magnus another progressive thinker had begun to establish himself, one who has influenced us more today: Roger Bacon (1214–1292). For Roger Bacon, as for many of today's scientists, the long-term memory of philosophical traditions was no longer essential. Rather, he argued that one had to be guided by mathematics and repeatable experiments that actively intervene in nature. Instead of a method of innovation based on the quest for wisdom of our ancestors, Roger Bacon wanted to lay the foundation for a new, all-encompassing *philosophia experimentalis*. He became the pioneer of a science program that does not consider nature as a system of values in which the new is to be harmoniously embedded, but in which it is regarded as a mathematically controllable field of experimentation. He predicted that this experimental mathematical science would lead us to the extension of life, the invention of flying machines, self-propelled carriages, and underwater moving instruments.[196] Hence he foresaw a lot of what we benefit from today. With his perspective, the modern concept of progress was born: the simple yet novel conceit that new = good.[197]

Of course, it still had to be proved that "new" really would be automatically "good." Philosophizing about it alone would not have cut the mustard. However, in the following decades and centuries, so many important inventions were made that the "new = good" idea began to appear more and more plausible. In 1269, the dry compass was invented, which would improve the efficiency and speed of transport and maritime trade.[198] Gunpowder arrived in 1359, meaning that European countries now possessed a military superiority that allowed them to spread across the globe and gain untold wealth through technical superiority. In the 14th century, the first wheel clocks came about, changing human rhythms and allowing for a sense of shared time, which again enabled more pliable forms of cooperation, planning and production. In 1458, book printing was invented, which greatly accelerated the spread of knowledge and education. By the end of the 16th century, European societies were so intoxicated with the power of new inventions that Roger's namesake, Francis Bacon, was able to conclude that "no empire or sect or star seems to have exercised a greater power and influence on human affairs than those mechanical things."[199]

Against this historical background, it is not surprising that our current thinking about progress is so deeply interwoven with new technologies and inventions, to the point where everything new automatically appears to us to be progressive. New services or products, new production processes, new markets, new organizational forms and manufacturing processes, new raw materials or substances – all this comes to mind when we think about the concept of progress today. We believe that our human well-being is nourished by the breast of invention. Any digital innovation that gets a wave of approval from the "invisible hand of the market" seems to be automatically, unquestionably celebrated as a contribution to humanity. The progression of the individual in the ancient sense seems to have faded into obscurity.

5.1.2 Progress Today Implies Old = Bad

While Roger Bacon is said to have still appreciated the wisdom of Aristotle, Francis Bacon (1561–1626), who lived 300 years later, not only refused to cultivate traditional forms of philosophical knowledge but even felt that they were a hindrance to progress. This notion was shared by many of his contemporaries. René Descartes (1596–1650) wrote that all those who followed Aristotle were like ivy, trying to climb higher than the tree that bore them. In the end, he contended, they would only grow more ignorant than they would have been without their extensive studies of antiquity. Those who knew least about the term "philosophy" would most likely understand the truth.[200] With these statements, an idea came into the world that is just as characteristic for our thinking today as the equation new = good; namely, the tendency to regard everything old or existing as relatively worse than the new thing replacing it.

In short: old = bad. Against this backdrop, what followed was a systematic rupture with the educational concept of the past. It was at the Royal Society, founded in 1660 and the meeting place of aspiring European scholars of the time, that the debate intensified. Defenders of ancient philosophy argued that the quest for insight from thinkers such as Aristotle had opened up a greater and deeper knowledge than the Royal Society and the entire current age. Proponents of the new belief in progress shot back across the bow, arguing that the Royal Society had achieved more in the few years of its existence than the Aristotelians in the entire previous 2,000 years before. The present age, they pointed out, had achieved many inventions – the microscope, telescope, thermometer, barometer, air pump – as well as making great leaps in geographic knowledge.[201]

The fact that apples were compared to pears in these intellectual battles should be easier to understand from today's perspective. It was simply a more humanistic approach to the sciences on one side that was contrasted with a more technically oriented way of thinking on the other. That these may actually be two complementary methods of "making progress" was dismissed. In the founding phase of modern science, the warriors at the vanguard did not try to find the golden mean for balancing the arguments. One position had to win. And if we look at our classical-modern thought patterns today, we know who wears the victor's laurels: the belief in the good of all that is new, the inferiority of the old, and the one-sided association of the concept of progress with new stuff and forms of scientific mastery of nature. Whatever antiquity or medieval thinkers may have found out about the need to guide people to moderation, their concepts of holistic education have been increasingly sequestered, hidden away in a historically tolerated yet ultimately neglected epochal drawer since the 17th century.

5.1.3 Progress Can Be Predicted

In the interceding years (between then and now), the future has been thought of as a haven of progress thanks to unprecedented technical inventions and scientific innovation. New = good and old = bad dominated the thinking of the elites. But not only that: this kind of progress was even considered to be God-given. Philosophers and scholars of the 17th century interpreted the pursuit of technological progress as a way to restore the "mastery of nature" lost in the Biblical fall of man from paradise. As late as the 19th century, Thomas Edison, the inventor of the light bulb, wrote about God's constant presence in his thoughts.[202] For religious people in Europe and America, this link between technological progress and God's will to restore paradise should not be understimated. When something is changed, the initial animal reflex of us humans is to balk at it. However, the belief that God himself might want the change seems to have worked as a key motivation for people to participate in them.

Crucial to the association of technological achievement with the reconquest of paradise was the idea of a *continuous* – if not linear – progression of the whole of humanity. On society's way to paradise, the later-born are smarter than their forefathers. This idea of continuity, grounded in a constant, linear progression toward paradise, suggested that the future might also be conceived as predictable. It was assumed that the future must be better when traveling in the direction of heavenly perfection. And so, in 1794, the French scholar Marquis de Condorcet wrote that man's history of progress is "the basis of his experience of the past," such that he is now able to, "with high probability, forecast the events of the future."[203] With these reflections, Condorcet became one of the first thinkers to express the idea of a calculable or predictable progress fed by the knowledge of the past. Our culture of prediction and growth models was born.

5.2 Seeing the World as a Predictable Model

While in Silicon Valley, I was recommended a science fiction novel by Isaac Asimov called *The Foundation Trilogy*. A central element of this work is a science called "psychohistory," which uses mathematical-statistical methods to calculate the future using behavioral data.[204] Given the looming demise of an interstellar empire, the inventor of psychohistory, Hari Seldon, predicts in this story how the galaxy's evolution will only continue if knowledge is saved on some outer planet, later called the Foundation. Many decisions are then wisely taken by the Foundation members on the basis of his scientific divinations.

The Foundation Trilogy, and indeed the whole field of science fiction, plays a central role in today's technological development. Almost all of the developers I have met in the past 25 years read science fiction, and almost everyone is aware of the concept of psychohistory. The movie *Minority Report* is also about psychohistory: cyborg-like beings (called "precogs" in the movie) predict where the next criminal offense will take place, whereupon a police unit is sent to the scene to prevent crime. Modern types of software called "predictive policing" go in the same direction. Police operations in many countries are how increasingly planned with software that uses algorithms to predict where crimes are expected, and police patrols can then proactively go there. This is true in the USA just as much as in Europe.[205] In fact, the tongue-in-cheek "Big Brother Award," awarded to Organizations that threaten personal privacy, was given in 2018 to the regional German government of Hesse that was pushing to take things even further: there the idea was mooted that police patrols should not only drive proactively into purportedly "dangerous" residential areas but that they should also be allowed to pre-emptively arrest citizens with sufficient statistical probability of committing a crime and observe them for three months.[206] Condorcet's idea of the calculability of the future seems to have arrived fully in our everyday lives as citizens.

But what is it that justifies our trust in such predictions? Can science fiction novels and Condorcet's journey to paradise really justify our deep faith in them? Not really. What was also needed was humanity's growing faith in the controllability of nature based on relaible models. And this faith was even born 300 years before Condorcet. It was in the early 15th century that sculptor and architect Filippo Brunelleschi posed in front of the baptistry of Florence, confronting his townspeople with an unusual experiment. He had discovered that by using spatial depth modeling, which we now call the "central perspective," we can draw images in such a way that they look fully realistic. With the central perspective, he reproduced the Florence baptistry in a highly realistic manner. And now he asked his fellow citizens to see whether his mathematically precise reproduction of reality could be the actual building. The reproduction he had painted was accessible through a small peephole. A citizen who peered through the peephole and looked into a mirror seeing the reflection of the painted baptistry got the impression of looking at the baptistry itself. When Brunelleschi then set aside the picture and the mirror and compared the real edifice to the mirror image, he illustrated that his model and the reality were amazingly similar.

Brunelleschi's experiment put forth an idea that is fundamental to our thinking today: we believe that we can capture and reproduce reality with our self-created models and images, such as is experienced in central perspective drawing.[207] The very idea must have had a highly reassuring effect on the people of that time, who lived very much at the mercy of nature. The great plague epidemics (1346 and 1353) were a recent occurrence, wiping out as much as a third of the European population. The idea that one could now capture, mimic, and perhaps even explain reality in a model using a simple method like this to gain more control over the vagaries of nature was a visionary promise. Metaphorically, it hinted at the possibility of decoding the mystery of the world and making nature more predictable – if not controllable.

Fig. 5.1a: The Cathedral of Florence.

Fig. 5.1b: Brunelleschi's experiment.

5.2.1 The Ultimate Belief in Controlling the World Through Models

Brunelleschi's central perspective experiment was designed to capture spatial reality in a model. Condorcet foregrounded the idea of temporal forecasting models. Both have in common that they tried to capture the unpredictable and multilayered essence of nature and to render it predictable and controllable – the guiding schema of today's science. Almost every technical innovation nowadays relies on models to function. But are we relying on a limited resource?

In Chapter 2, I showed how we can try to capture the coziness of an apartment in a function. However, as an AI developer, we first need a model that allows us to teach the phenomenon of coziness to a machine. What does such a model look like? In order to develop a useful one, we first dissect the phenomenon to be reproduced – in this case, the somewhat abstract idea of "coziness." Then, we break it down into its individual components and properties. For example, we could say that coziness has something to do with light; maybe a golden or a muted light. And maybe we could add certain object types; previously, I mentioned a fireplace, for example. Perhaps comfortable armchairs and blankets also fit the bill. In short, as developers we have to think with a good dose of imagination about the most important components of a phenomenon to be functionally reproduced. And then we try to capture these in an algorithm, or train them into an LLM. We insert the individual components (or strenghten codes) in a suitable mathematical formula (or network) and have a model: in this case, for example, coziness = light f (golden, dimmed) + objects f (fireplace, blankets and armchairs).[208] Such a model can then subsequently be checked with the help of experiment surveys and statistical evaluation methods which confirm (or refute) with a certain probability that the coziness has been aptly captured.[209]

But can we really capture the phenomenon of coziness with such a simplified model, or even a much more complex AI model? Maybe there are quite cozy rooms that are bright and airy, without a fireplace, armchairs and blankets? Without a doubt: just think of a Swedish dining room in the golden hour of morning. From this (and many other examples), we should actually deduce that models, like all algorithms based on them, are always at risk of being incomplete. They cannot do full justice to the nonidentical repetitions of realities of life and the invisible ideas inherent in human values (such as coziness). Algorithms lack access to the singularitas of our world, interspersed and interwoven as it is with the invisible *Gestalten* through which values infuse our world.

Why is it, then, still important to discuss modern model-based thinking in such detail? Well, in 15th century Europe, the conviction was born that the whole world could be apprehended in models. On this basis, we presume today to be able to grasp an abstract idea in a model; say, crime, willingness to work or even humanity. It is believed that models could grasp the complexities of all natural phenomena, humanity included, thereby paving the way to a paradisiacal future in which humanity held the key of knowledge in its hand. This unshakeable historical belief governs us today, causing us to set great store by the often erroneous market forecasts of our research institutes. The assumption that models can adequately reproduce our world despite their abstractions and brutal simplifications is what leads us to blindly trust that an AI can reliably figure out the coziness of a room, if not today, then in some paradisiacal future. And we even believe that the model-based machine is more accurate than the human being, whose estimation of coziness we are skeptical about, because people act intuitively. Intuition, however, is believed to be neither rational and controllable nor modelable, and therefore not good. The belief in the supposed scientific nature of incomplete models is so strong that we habitually assume that machines must always be more correct than we are. This assumption may be problematic though.

I point to the word "belief" here. Objectively, no model, no AI and thus no machine is able to capture the singularitas of life; they are not even able to truly portray simple value phenomena such as coziness, let alone complex value configurations as they exist in reality.[210] And yet our belief in models and such systems is unshakeable. It has been drummed into us for 500 years.

5.2.2 When Models Become Dangerous

Today, we are so convinced of the success of models that we have even started to subjugate reality to them. What does not exist in the model, we believe, does not exist in reality either. And what exists in the model, of course, must exist in reality. Such thinking has an impact on our political decisions, as we can see from the increased use of predictive policing.

But consider another current example that is used for model-based policy making: the age pyramid model, which shows aging for many industrialized nations. The age pyramid is predicting that in the medium term, many European countries are barreling toward a crisis: there are too few young people to support the disproportionately large number of retired people. Of course, one can prove the existence of the age pyramid and its predictive power, if – as is common for such models – the reality of our population growth is reduced to *domestic* births, while ignoring immigration. And indeed, the predictive power of the age pyramid is perfect in those industries where native-born citizens are a preferred source of labor or in those countries where cultural and linguistic barriers may deter immigrants (e.g., Japan). Nevertheless, the model is deceptive because in reality, the population growth in many European nations is not driven by births anymore, but by the migration of peoples, that is, by factors outside the age pyramid model. Naturally, there will always be net migration from countries with weaker economies to those with stronger ones. At the Vienna University of Economics and Business in 2024, 38% of students were international.[211] Of my 6 permanent employees, only one was born in Austria. I expect few of them will return to their original homeland. Even before the refugee crisis in 2016, the population of Germany was growing by several hundred thousand per year.[212] Many of these migrants compete for jobs in the labor market and many integrated workers today pay heavily into our social security funds and are appreciated colleagues.

Why is a realistic balancing of a model, such as the age pyramid, important for our progress-driven policies? This became clear to me when a board member of Germany's biggest trade union reassured me that he was not worried about the increasing automation of production facilities in Germany since the age pyramid would soon lead to a natural shortage of employees. Such a statement is a frightening example of how much model-based thinking is internalized by our elites. Does the representative of just under 2.1 million union members not consider financially precarious refugees and migrants, many of whom are in need of just the kind of job that is now automated?[213] While the comrades who pay membership fees speak of "factories of fear" where no one works anymore due to robots, their representative resides in a magnificent glass palace in downtown Frankfurt and studies models, to which he pays more attention than the real-life conversations with the people on the ground that he is supposed to represent.

But should we consider the naive faith in models to be only a problem of the elites? Definitely not! In fact, we are all blithely traveling together along towards historical dead end. On a private level, we carry models with great enthusiasm and sincerity in our daily lives as well. For example, instead of looking at the shape of our body to find out whether we have become fat or thin, we look at our scales to see what shape we are in. But this downward gaze does not by any means tell us the full truth about our weight, because muscles, menstruation, the time of day of the measurement, and a pitcher of beer the night before the weigh-in are decisive factors that can strongly influence that little number on the display. But we ignore this, of course. We are annoyed, or self-deceivingly jubilant that today we have gained or lost a pound. And we have fun with it! We also count our friends and our likes on Facebook, Tiktok and LinkedIn to fool ourselves that we are popular or to lament that we are not. We almost only look into the Brunelleschian mirrors of our inventions and models and believe that we see the truth about ourselves in these self-deceiving reflections of reality.

5.3 The World as an Object of Design

Of course, as a scientist and world designer, I have to admit that I, too like to build models and do so constantly in order to be able to publish academically. Models are for a curious, knowledge-hungry person like compasses that help point the way to understanding the world. However, more models and inventions are born these days than we could ever seriously need. Every year, nearly 30,000 young people in Germany receive a PhD, of which all are supposed to contrive a model or refine an existing one.[214] This is just one aspect contributing to an explosion of ideas of how the world supposedly works.

However, the story does not end when thousands of models, technologies and world ideas are on the table: once you've invented a model or proven it with some statistics, or you've made a technical invention, then there inevitably comes the point at which third-party research funds ask, "So what? How relevant is your model or your new technology? Can you prove that you can disseminate it in reality? How much profit can it make for us?"

And this gives me the chance to share a quirky anecdote about QR-coded biscuits, whose development I had the chance to marvel at a few years ago at the German Center for Artificial Intelligence (DFKI). A doctoral student had developed a technique that allowed QR codes to be baked into cookies.[215] He told us that it would be nice to decorate cookies with QR codes instead of some colorful sprinkles. He imagined that one could then gift his or her beloved QR code cookies, and the recipients could then use the QR code reader on their cell phone to navigate to a website, where a sweet message would await them before the meal. I do not know what you, dear readers, think about this invention, but I can assure you on my part that it was only my courtesy that did not make me laugh at this well-intentioned but also totally bizarre idea. Who on earth wants unsightly QR codes on biscuits? And who wants to look at digital messages instead of eating the biscuit right away? Such realism is not welcome in scientific appreciation, of course. What counts is only that the biscuit somehow sells and that you published a peer-reviewed article based on your idea.

Funnily enough, on this occasion, the "invisible hand of the market" deemed the biscuits to have passed muster: the QR code cookies were in fact available to buy. But the example, with all its absurdity, reveals a recurrent pattern of modern scientific thinking, which I call "The Versailles Model of Progress" (see Fig. 5.2). Designers, in their roles as inventors, researchers, start-uppers and scientists, conceive new ideas and models – and then bluntly redesign the world according to their more or less fantastic ideas. The world, for them, is a dollhouse to play with, a house they can rearrange as they see fit. Thanks to endless money markets, they then even find ways to push their phantasmagorical ideas onto society. But they often ignore that the brainchild you bring into the world may be far from the lived values sought in reality.

5.3.1 Parks of Versailles Versus English Gardens

The idea that the world should be allowed to artificially expand like a baking loaf, as a steady process, and develop according to self-made values, does not occur to most world designers. For them the only relevant axiom, as we have discussed, is that new = good and old = bad. And where the market and its users are not yet ready for their creations, they can count on a marketing culture that artificially generates demand with a lot of marketing money, or asserts sales power through lobbying.

Fig. 5.2: The Gardens of Versailles (own depiction).

I call this the "The Versailles model of progress" because if we take a walk at the park of Versailles, we will find that this park is a perfect design piece that lacks any natural relaxation (Fig. 5.2). The park came about when Louis XIV had a marshy area drained, which the original inhabitants referred to as the "stinking pools." The king laid out a park in which each tree had to be trimmed for 20 to 30 years until it finally looked like a musical instrument. If you ever visit the park, you will walk on broad, symmetrical, dust-dry paths until you are completely exhausted, occasionally observing a palm that would never naturally thrive in the climate of Versailles. As early as the 17th century, a microclimate was artificially created that allowed asparagus and strawberries to be harvested in the winter – a tremendous innovation at the time. Louis XIV personally wrote a manual on the "manner to walk the gardens of Versailles." Just like many of our digital technologies, such places are hardly accessible to the visitor in any kind of intuitive way.

The Gardens of Versailles symbolize a form of progress that sees the natural world as an object of manipulation for the powerful designer. At the starting point is the mostly context-independent model of the designer; this model is then imposed upon nature by hook or by crook and at any cost. The designer, given his political or economic power, feels no need to inquire about realities, cultures or contextual values. He is not interested in the existing potentials of the local value-bearers such as the earth, the climate, the landscape or human needs. He is not interested in what *is*. Reality has no meaning for him because he assumes that you can trim it like boxwood.

The kind of "progress" that Versailles calls for is perfectly controlled and imprinted with a sublime brutality that smoothes over the "imperfections" of reality. As an expression of absolutistic thinking, the Versailles model of progress was copied and admired all over the world. But it also represents a form of progress that is infinitely laborious to create and sustain. Transporting every palm – each weighing more than 40kg – into a palm house every winter and then putting it back in the park every spring means a lot of work for the gardeners, not to mention all the pruning each tree requires. The designer and his executives, locked in a constant struggle against the resistance of nature, desperately waste their energy on the fulfillment of their beloved brainchild. I consciously use the verb "wasting" because it is quite certain that there are other ways to create an elaborate, beautiful and intoxicating garden. Take, for instance, an English garden (Fig. 5.3).

Fig. 5.3: An English landscape garden (own depiction).

For a start, an English landscape gardener would probably never create a dust-dry designer landscape in a place where marshland prevails. Instead, he or she would analyze the existing conditions and value potentials of the context and consider which water courses and which plants could thrive in such an environment. They would, of course, tame and channel the water using state-of-the-art methods, working with the existing conditions to cultivate the landscape with care, working tentatively where the gradient advances. They would pay attention to natural enclosures, instead of imposing a foreign material on the landscape. In addition, by allowing trees and flowers to thrive in their ancestral landscape, the gardener would help them to attain healthy, beautiful growth. What is more, this approach would also save energy, which others waste in the creation of their blinkered contrivances.

In an English garden, much is as well thought out and skillful as in a Versailles park. The educated classical reason, which is based on trained value-intuition and knowledge of nature, receives more attention here. Again, the space and growth of a plant are not left to nature itself. However, as with Albertus Magnus's concept of progress, this arrangement takes place in harmony with the prevailing environment, and not just with regard to an abstract design model that has to be fulfilled. Things are intuitively and knowingly related to each other and coordinated, and technological aids are put to this service. It is not about perfect control, but merely directing and giving space to the living, to allow it to develop on its own. The result of this tuning is harmony in the eye of the beholder.

English gardening represents the kind of progress I want to propose in this book. I believe that the inventors and scientists of the world – and especially the IT industry – should work on this form of English garden progress rather than cultivating complex contrivances without appreciating the given context; IT innovators should solve the problems that are there and not invent things that nobody needs. The Versailles model, in my view, does not promise true progress for the future of mankind, but represents only wasted effort and regression.

5.3.2 Progress of Humanity Versus Progress of the Individual

In the 18th century, another formative mentality emerged that needs to be understood to see why many have become skeptical about progress. Why do so many of us not appreciate the "progress" offered by Silicon Valley? In fact, thinkers in the 17th and 18th centuries not only moved away from nature, which was increasingly perceived as something to be "managed," but they also moved away from humanity itself. In ancient times, philosophical thought was guided by the aspiration for the good life and action of the individual. Enlightenment philosophers, on the other hand, increasingly overlooked the fate of individuals or local communities and instead considered grander and more ambitious scales, such as the "education of the human race" – the title of a major work by Gotthold Ephraim Lessing in 1780. Like Condorcet, they reflected on "a revolution in the destiny of the human race at full scale."[216] Kant recommended in the categorical imperative that we should focus on "*humanity*, in your own person as well as in the person of any other" and established thinking about "universal" reason.[217] The naturalist and Jacobin Johann Georg Forster, meanwhile, described in a guide to a future history of mankind how the "local, special, uniqueness" of human beings should be transformed into a "universality." We find ourselves on a path where we are "as it were, an idealized people, abstracted from the whole of the human race." And as individuals, we are little more than "representatives of the whole species."[218]

In this vein, Kant established the idea that the human race would not automatically benefit from progress. Rather, every individual was obliged to contribute to it by using his reason "critically." The individual is called upon to put her life into the service of overall social progress. This is where the free individual comes into play: we are free to decide what we do – provided we follow the abstract rules and duties described in Kant's most famous concept, *the categorical imperative*. Firstly, however, we have to inform ourselves about what is desirable for humanity. Thus, to some extent, the history of progress in Kant seems disconnected from our immediate context of local engagement where we as responsible citizens resonate with our community in context. Much like the classical thinkers of British liberalism, Kant consolidated a conception of the human in general that places our thinking at the service of an abstract "universal" ideal of progress that we have the duty to live up to.

To this day, the thinking of the liberal elites of Europe and America is shaped by this abstract ideal of progress.[219] In all areas of life, and especially in the use of new technologies, we feel the progressive and apparently universal obligation to "stay on the ball." We strive to impress others through abstract education, awareness of the latest trends and innovative action, and do not realize that we basically function as hamsters in the progress-wheels of others' pet ideas. Even the eldest within current societies are trying to deal with mobile apps, social networks and AI – despite, perhaps, some internal resistance – not because they need it or want it, but because they think that "progressiveness" is expected of them; that in some sense, they have to "move with the times." Others feel "up to date" when they sink into the perpetual cycle of LinkedIn, TikTok or Facebook. And all these have in common that the motivation for their actions often comes externally: from society, and not from their innermost wishes.

This passive, duty-driven, and mainstream position of "technology acceptance" in which most people today find themselves in the face of progress, does not do justice to the creative and intellectual potential of the individual. It is a strangely alienated role that does not fit the image of the enlightened, autonomous, self-determined individual. Could it be that the modern age's image of humanity to which we have submitted is not as rational as Kant made us believe?

5.4 Transhumanism: An Ideology of Lovelessness

Passivity vis-à-vis technological progress could be tolerable if we were right to trust in progress being to the benefit of humanity. Not everyone is an active designer by nature, or feels called to define the digital future. But at the very least, humanity must be able to trust that those who advance technology in our name will do so responsibly, modestly, and with a sincere wish to improve the lot of mankind. However, this is less and less the case. Instead, it is striking how new technologies are increasingly being used to replace us rather than strengthen us, driving us into dependencies and alienating us from each other.

As I have already described, computer science has not yet had its equivalent of the "organic revolution" in the food industry. IT systems are still being developed without much foresight. Even though the HCI community has made consistent efforts for decades to make software more user-friendly and also tries to put people's well-being back at the center of development processes, many system developers are working with a questionable understanding of mankind. Their perspective finds its expression in some of the more uncharitable names they find for us humans and our bodies. For example, "DAU" ("dumbest assumable user"), or references to us having "meat sticks" (rather than fingers).

A particularly impressive experience illustrating the misanthropic thinking of tech elites was showcased a few years ago at a conference in Switzerland. The day of the conference began with a talk by a professor of computer science at the Swiss Federal Institute of Technology in Zurich (ETH), who titled his presentation slides as follows: On the first slide it read: "0:1 against humanity," next slide "0:2 against humanity," then 0:3, and so on. I wondered, who is actually playing against humans here? From the point of view of the smugly self-convinced scientist, it was stupid humanity that would lose against artificial intelligence. The scientist then feverishly described the vision of AI that he would be dedicated to in his work, presenting scenes from the movie *Her*.[220] This movie is about a middle-aged man named Theodore who divorces, before falling in love with a speech assistant called Samantha. Through headphones he can constantly hear Samantha, an AI system that is similar to the Alexa system of Amazon, but with the response quality of ChatGPT or Claude. In the film, Theodore lives largely in isolation from his natural environment and finds himself in a state of increasing dependence on Samantha, with which/whom he falls in love. Eventually, he is completely delivered over to the manipulation ability of the language system. He panics if he does not reach it/her; he is no longer himself and is ultimately left deeply disillusioned when he finally learns that Samantha has served him at the same time as 8,316 other users – in other words, she has been serially unfaithful to him.

The way in which the Zurich scientist defined "intelligence" as surpassing all human consciousness was a bit simplistic: he obviously equated the term "intelligence" with a computer's ability to beat a human player in a closed, rule-based environment, as is the case in a chess game. This has little to do, of course, with intelligence in the human sense, which is precisely characterized by the integration of multilayered cognitive and emotional *understanding of relevancies* in open, non-rule-based, cross-context situations. But apart from his limited understanding of intelligence, he also seemed fascinated by the ability of Samantha to become so deeply intimate with her owner; in Samantha's ability to simulate a nonexistent love seamlessly. The scientific jargon for such simulations of emotion is "Affective Computing."

Does the ETH scientist not notice that he is working on a very sad, completely dehumanized and – more importantly – ethically extremely questionable future? A transhumanistic future? A future in which people act like miserable dishcloths and in which they are sad, alone and dependent on technology and artificial love; a future that manipulates us and fools us with simulated AI feelings? The scientist's euphoria stemmed from the notion that such emotionally charged conversations between software agents and humans are technically feasible, but that technical feasibility seemed charged to the point where it fully blinded him to the human consequences of the technology. And this one ETH professor is no exception. Many scientists believe that the ultimate end of AI perfection is to build as human-like as possible and thereby foster human-machine relationships as they are portrayed in the movie *Her*. They are also supported by industry, because humanlike AI characters are more likely to bond with users and increase human trust, which again drives business models. On the other hand, there are scientists like myself who are deeply concerned about the effects of such new forms of bonding as well as the lack of safety that is related to this kind of AI use. Some users of humanlike AIs have already killed themselves to be closer to their beloved software agents or followed the most ridiculous advice made up by their large-language-models (LLMs).[221] Safety issues are of course a concern shared by the chatbot industry, which has published international research calls seeking scientific advice on fixing bots' "alignment problem" with humans. But the outlook seems dim as to whether reliable solutions to the collateral damage can be found: software agents simply aren't capable of recognizing complex values, it seems,[222] and this might even be an unsolvable technical problem because "relevance realization" seems to require aliveness, a trait software agents do not possess.[223] But even if these technical challenges were somehow overcome to a sufficient degree, the question is still open as to what human-AI bonding does to humans' development as persons, to social communities, to democratic and convivial behavior, to people's language, knowledge, intelligence and psychological stability.

What is not in question, but seems to occur with increasing frequency in classrooms, is the phenomenon of pupils challenging their human teachers, believing more in LLMs as an information source than the person entrusted with their education. Günther Anders' vision of Man as an "antiquated" being has not only embedded itself deeply into the minds of today's scientific elites but is becoming reality.[224] Many people perceive humans and their nature as suboptimal, stupid and unhappy "systems" from which AI saves us. They believe that mankind must be improved and, according to the Versailles style, trimmed like a boxwood with technology, so that the world becomes as straightforward as the dust-dry sand roads of an absolutist park. They have no respect for human dignity as it is, with all its endearing weaknesses, disabilities and uniqueness that make up the diversity that constitutes a true blissful state. In their eyes, we no longer need human friends, lovers and counselors: we need "unbiased" software assistants.

5.4.1 Two Contradictory Ideas on Humankind

When I talk to the people that espouse such ideas in more detail and ask them why they think the way they do, and where all the AI, big data, implanted chips and blockchains should bring us, I always get the same answer: we need these technologies because only they can solve the problems of humanity. Man is ultimately fallible, they argue. History, they insist, has shown that humans are too weak, unpredictable, selfish, opportunistic, fear-driven, emotional, irrational, too slow; too inaccurate, inefficient and prone to disease, whims and violent conflicts. Machines are simply more predictable, more durable and more intelligent than humans; they should therefore be used to improve people and to control our fallibilities. One ought, they explain, to at least steer people in the right direction, or even better, simply technologically improve them. The latter, in technical jargon, is referred to as "enhancing," with the clear implication that it makes us "better." It strikes me that one must possess a frighteningly contemptuous and negative idea of humanity to embrace such arguments.

I must confess that it is an alienating experience to hear my contemporaries speak like that, because it is not only my own experience that tells me that this idea of humanity cannot be right. The psychological, sociological, philosophical and economic research of the last two decades shows us a much more positive picture of men and women. With all the weaknesses and confusions that we humans might have, we are incredibly beautiful, powerful and good at the same time. We are social beings who, beyond our own survival, seek love and friendship. As some authors have recently shown, we are much more "homo puppies" that survived due to cooperation and conviviality than we are beasts surviving only on competitive advantage, or survival of the fittest.[225] In our lives, we can combine and articulate our rational and emotional thinking and perception to create the greatest poetry and art. We thrive on stories, symbols and metaphors. Notwithstanding the highly complex and undecipherable nature of our bodies, nature ensures that we are healthy for most of our lives. We are often able to produce peak physical performance at a sublime level that is hard to believe. We wish to live peacefully and contentedly, and many of us recognize our moods and weaknesses with age. If we look closely, we have excellent intelligence that allows us to grasp the reality of our organic and social world. Above all, we have the sensory apparatus to capture the world of values that infuses our lives with meaning and purpose. As I have already described in detail, digital machines cannot even begin to project into such dimensions. Most people are not predominantly selfish. On the contrary, they dedicate most of their lives to others, be it as parents, friends, spouses, corporate and government officials, or as volunteers. We have a natural tendency to favor the good and the beautiful, and we turn out to be pretty fair in most economic experiments.

To dissuade ourselves from this knowledge about ourselves and our trust in humanity, and to see ourselves in the pitiful light the IT industry likes to push us into, must stand as a nadir in our intellectual history. How could it happen that many question human nature to such an extent that they see progress in "enhancing" man, or even replacing our species with a new generation of cyborgs?[226] I must mention a few sources of this thinking so that we can see why so many perceive humans so negatively. Only when we realize that the unfavorable opinion of ourselves has been (and still is) a political program might we be able to say goodbye to this negative thinking.

5.4.2 Historic Sources for Negative Ideas on Humankind

Many believe that humanism as a spiritual premise inspired the Enlightenment. Humanism sketched a promising, positive image of man and his developmental abilities, but at the same time, a negative image of mankind was also established.

Undoubtedly, the roots of this negative perspective on ourselves are present in the way Western Christianity has taught and interpreted some core aspects of its faith since the onset of modernity. Although Christianity and Judaism are based on mercy and charity, and Christianity specifically established the idea of the ontological dignity of all people, the biblical salvation story also embraces the fallibility and weakness of human beings. The stories refer to the expulsion from paradise, to Sodom and Gomorrah, the struggle and envy between family members (Jacob and Esau), the thirst for power of the elites (pharaohs, pharisees, etc.), the betrayal of friends (through Peter and Judas), and the damned who "howl and gnash their teeth" (Matthew 25:30). It is no coincidence that such stories became more and more important at the beginning of modern times. Martin Luther (1483–1546) proclaimed that all people who did not study the Scriptures were lost. John Calvin (1509–1564), meanwhile, preached that mankind was divided into the few who are destined for heaven and the many who will go to hell. And for the Catholic Cornelius Jansen (1585–1638), it was a lost effort to seek God's salvation out of free volition: "The void awaits surely all of them that weave the wind."[227]

This cautionary, ambiguous, easily misinterpreted idea of humanity was accelerated by non-church thinkers. Thomas Hobbes (1588–1679) was particularly influential in this regard. In his work *Leviathan*, written in exile in France during the English Civil War, he attacked the prevailing Aristotelian notion that humans are "political" or "social" animals who cultivate their lives in communion with other human beings.[228] In the light of his impressions of the war, he characterized people instead as atomized individuals who were chiefly interested only in their own benefit and well-being. Based on these impressions is his idea that, in the light of our individual optimizations, we ultimately live in a "war of all against all" and everyone is only concerned with his own advantage. This is also the background of his realization: "Man is a wolf to man." In the supposedly "primordial" natural state, Hobbes said, there were no innate ties of nature in man. Desires, fear, egoism, competition and gluttony would cause even peaceful individuals to encounter each other with suspicion. This anarchic condition can only be overcome if people abide by laws, whose validity is enforced by an absolutist monarch. Of course, the absolutist rulers of the time gladly took up Hobbes' thoughts and contributed to their dissemination. This frankly archaic idea formed the beginnings of a mental bedrock that every economist learns today in his undergraduate studies: the idea of man as a private benefit-optimizer, as homo œconomicus.

Even John Locke (1632–1704), the father of liberalism, advocated the idea of a natural state of human beings as atomized individuals rather than as communal entities. By virtue of their nature, according to Locke, humans are born with a right to life, freedom and private property. In contrast to Hobbes, he believed in the ability of human beings to follow a moral code of their own accord; however, at the same time, he was also convinced that people tend to violate rules because of their individualistic impulses and self-interests. People, he suggested, live together in a state of "fragile peace." Therefore, it is necessary to pass laws and appoint judges who protect the moral order of the state of nature on behalf of the citizens.[229]

Finally, Jean-Jacques Rousseau (1712–1778) should be mentioned in this series of critical thinkers. Unlike Hobbes, it was Rousseau's conviction that man in the state of nature is a peaceful being, but at the same time, he was convinced that the merger of atomized individuals would inevitably lead to envy and hatred. Rousseau was skeptical about our spontaneous ability to live in community: "Let human society be admired as much as one wants; it will be no less true for it that it necessarily brings men to hate one another to the extent that their interests are at cross-purposes with one another, to render mutually to one another apparent services and in fact, do every evil imaginable to one another."[230] Rousseau's solution to this maelstrom of conflicting human desires, in addition to his ideas for an educational program, was a lawmaker who would serve the common good. This lawmaker should protect *the general interest* in all areas in which the individual and the economic system might fail.

The absolutist monarchies of the time were again amenable to this portrayal of humans. It could be used as a handy argument to build a hitherto nonexistent state apparatus. Reform programs and laws that cemented the powers of monarchs could now be legitimized under the spurious guise of protecting people from themselves. These monarchies may have mostly disappeared nowadays but the state apparatuses have remained, and with them also this questionable image of mankind, which is still useful today to justify the state structure and its law-producing machinery. If we combine Hobbes, Locke and Rousseau, we effectively arrive at the party-political thinking of today's democracies. We can then call, either in the Rousseau style of the "left," for a strong state that protects people from injustice and unequal treatment; or, in the Hobbesian bent of the "conservatives," for a lean but effective law and order. Both styles of government represent regimes that direct the supposedly natural war of all against all in orderly paths.

Ultimately, the difference between these viewpoints is negligible for my argument here. What is germane is that both sides have stopped trusting in the inherent goodness of humanity. Thus, ethical thinking and confidence in individuals has been outsourced to sophisticated legal systems. People have gotten used to not having to get along with each other; in case of doubt or conflict they can simply call their lawyer. The value-oriented struggle for a virtuous life – as well as good business conduct and technical invention – has no place within this mindset.[231] "Anything goes" as long as it is not prohibited by law. Morality is delegated to the legal departments within companies. And IT developers and entrepreneurs educated in this mindset are developing technology for a humanity one cannot trust.

With these developments in mind, it should come as no surprise that FoodIS executives cannot easily implement the beautiful value world that my students sketched out in their second homework (portrayed in Section 3.2). Management thinking is just as saturated with the negative idea of humanity as that of the fathers of liberalism on which our economy is based. We should therefore not be surprised if technological innovations are used to kill, check, rank and evaluate the presumably bad people. Companies like FoodIS are afraid that employees will abuse their freedoms if they are not monitored and sanctioned. They want control over their quality of service, having learned to assume that every courier and restaurant will instinctively not care about service quality. For safety reasons, they prefer to have their couriers monitored by GPS and customer rating systems. Soon, the company could move on to the Disney method described previously, where employees are mere hamsters in the wheel and have to run as fast as they can to earn enough brownie points to keep their precarious jobs.

5.4.3 On the Loveless Ideas of Transhumanists

At the sharp end of this history of progress is transhumanism, which has today cast a spell over many tech elite circles.[232] In transhumanism, the negative idea of humanity condenses into an ideology of lovelessness, in which man is simply regarded as a suboptimal, obsolescent model to be phased out. In complete oblivion of history (Nazi eugenics date back a mere 80 years), transhumanists are openly thinking not only about optimizing and controlling people through machines but also experimenting with their bodies in order to improve, renew or revive them to bring a supposedly higher level of evolution.[233] Ideas such as the proactive replacement of biological organs, e.g., our hearts, by supposedly more suitable computerized carrier materials, are openly discussed.[234] Technology is believed to be able to improve healthy brains with external digital memory components, for example, by inserting brain implants to improve our memory.[235] Some transhumanist "scientists" even have the idea of simply scanning our brains and uploading all existing information to a computer. They call this process "uploading."[236] Google's product director, Ray Kurzweil, even declares outright that uploading processes would be "more efficient" if one had the ability to destroy the physical human brain while scanning it.[237] This type of "research" is currently being done on animals, at institutions such as Yale University, where people have been trying to bring the brains of hundreds of slaughtered pigs back to life.[238]

Transhumanists seem unable to comprehend that there is no information in our brain that can simply be scanned as if data was being copied from a hard drive.[239] They also seem to have little consideration for what "destroying" the brain of a living being actually means: it means to tamper with the sacredness of our bodies and kill what is still alive. Transhumanists imagine humans as a computer, where you can drag, drop and delete as you please. Clearly, far from many current scientific insights of the mind and body as a whole system, they seem to believe that the brain is our hardware and our minds are software running on it.[240] The brain is the processing unit in this concept, and the rest of the body supplies it with energy, so to speak, but otherwise has no really relevant functions for our thinking. Deeply rooted in the late 17th century image of humanity, contemporary transhumanists still believe that cognitive thinking alone determines being (I think, therefore I am) and then assume that this thinking takes place in the brain. Many, therefore, find the rest of the body superfluous; yes, even wanting to leave this outdated shell behind.[241]

Reflecting on transhumanist perspectives, the best strategy might be to remember the saying often attributed to Einstein: "You have to make things as simple as possible. But not simpler." For transhumanists, the world of values is irrelevant, since intuition and feelings are disposable and the body is unimportant. Even the singularitas of human consciousness, perception and being, and thus human temporality (i.e., the coming together of past, present and future) are either unknown to them or simply neglected as model variables.[242] Thus, transhumanism excludes much of what distinguishes humanity from machines: the complementing of our thinking through feelings and intuitions that come from the lived-body; our ability to experience the world as a space of values and to form extended integrals with our environments; our temporality and singularitas.

How is it, then, that transhumanists can nevertheless score points with ostensibly intelligent people and credibly spread this ideology of lovelessness, even among politicians and businesspeople? How can it be that the transhumanist ideology is currently even elevated to trend-setting knowledge promoted in textbooks for pupils?[243] As I have already described above, simplistic models are appealing because they are so easy to understand and calculable, and therefore still provide much of the bread and butter of modern sciences. And as long as a model, as simple as it may be, is not refuted, it is considered valid. Negative ideas of humanity enshrined in models such as the homo œconomicus still dominate its economics education and research so that transhumanists have good turf to thrive on. The value-oriented, emotionally intelligent thinking of human beings is regarded as unworthy of trust, as unscientific, or as a "highly paid personal opinion" (HIPPO for short).

As a result of their mainstream scientific positioning, transhumanists can play a very professional role as world designers. They accord themselves the right to make the world their own, just as Louis XIV had his boxwood trees trimmed without really looking at local realities and human destinies. If something goes wrong – if an Uber car kills a pedestrian or an LLM drives individuals into suicide – then that's ultimately "collateral damage" incurred in the interest of humanity's progress. There is only one red line for transhumanists: adhere more or less to existing laws. But these are not sufficiently available in the world of technology – and internationally hardly enforceable anyway. If worst comes to worst, big-enough tech companies can play ping pong with the courts,[244] or even outsource questionable, borderline practices to "free cities" or "international waters."[245]

With their narratives, transhumanists have succeeded in positioning themselves as best-selling scientists. Even though they seem to be less successful in the established higher education system with its still respectable appointment procedures, the funds of the IT industry have helped them to found their own "universities" (such as the "Singularity University"),[246] or at least lavishly equip industry-think tanks at elite universities, under whose honorable flag they then proceed to sail. The deep pockets of the IT industry finance this kind of progressive movement, which is partly academic in nature, and partly has quasi-religious ambitions.[247] And even though reputed universities are slowly starting to wake up to the abuse of their good names, many people still, for want of better knowledge, believe in the transhumanist visions.[248] We can only counter this dangerous development in democratic countries in one way: we must develop brighter, more thoughtful and more human-friendly alternative visions that can counter this dark and dangerous ideology. Then at least the economy, society and politics are presented with more options and can decide which way they want to go. But such a scenario is not easy to realize because, as French philosopher Simone Weil (1909–1943) put it once, "Imaginary evil is romantic and varied: . . . Imaginary good is boring: real good is always new, marvellous, intoxicating."[249]

6 Digital Ethics in Practice

When this book was first published in 2019, those who listened to the world of technology and AI heard a lot of enthusiasm and determination to integrate human values into AI design. The world's largest and oldest engineering association, the Institute of Electrical and Electronics Engineers (IEEE),[250] had established the biggest ever initiative for "ethical system design"[251] and prepared a dozen standards that were envisioned to help IT innovators build ethical and value-based AI systems. Over a period of seven years (starting in 2016), the association mobilized more than 1,000 experts worldwide in a bottom-up process to find out how to make technical systems more human-friendly, child-safe, privacy-protecting, etc. IEEE promoted a relevant list of value principles, which was supposed to give its roughly 486,000 members orientation as to the development of value-based technology.[252] I personally co-chaired and drafted the development of the IEEE 7000 standard, which in 2021 was the first ever globally standardized model process for ethical system design. The standard outlines how exactly we can embed value dispositions into IT systems so that human and societal risks are minimized. IEEE 7000, which was adopted by ISO in 2022 as ISO/IEC/IEEE 24748-7000 embraces high intrinsic values as core technological development goals, such as those endorsed in the Universal Declaration of Human Rights. It works the same way I made my students work: asking the same three ethical questions at the outset of an innovation project and then deriving system requirements for engineers and product managers.

IEEE did not, however, remain the only global institution publishing guidelines for ethical AI and system design. The second largest professional organization relevant in computer science, the Association of Computing Machinery (ACM), also revised its ethical guidelines for members. The OECD (in 2019 and 2024) put out a catalog of expectations on trustworthy and human-friendly AI.[253] It contained the value-based principles of accountability, robustness, security and safety, as well as transparency. Large value-chunks like a call for well-being-related values or human-rights enshrined values (including fairness and privacy), were put forth. In a 2019 summary article, the journal *Nature* reported how 84 institutions had made organizational commitments to these or similar AI value principles.[254] Finally, the United Nations worked extensively between 2022 and 2024 on a global governance framework for AI that the organization hopes can support all of these calls for a more ethical and value-based system landscape.[255]

Taken together, we see that there is a willingness to embrace Value-based-Engineering in the age of AI. Progress is perceived by many wise experts to only be

 | https://doi.org/10.1515/9783111677736-006

achieved if certain baseline value commitments are met, which is also the claim and call of this book. However, can we expect from these agreements that value-based technology is really around the corner? Is the money being put where the mouth is? And can we actually technically achieve what is right in a reasonable time frame with reasonable effort? I am not so sure.

6.1 Challenges to Ethical AI

If we take the value principle lists seriously, then many of today's IT systems, including AI systems are still utterly unethical. Consider the Disney laundry system described in Chapter 3, which undermines the freedom, dignity and health of workers; or the software used by police for predictive policing, which could undermine the privacy, equality, freedom and safety of people. Even addictive systems, like social media, anthropomorphic AI chat systems or the many gaming worlds are, considering the mental and physical health effects as well as well-being of their users, seriously questionable technologies. The many value frameworks are not met by most of today's digital services or AI offerings, though they were put forth years ago.

The reason for this sobering reality is that system developers today design their whole systems in a non-value-based way. Value-driven engineering requires upfront planning and ethical requirements engineering before one can actually start developing a system. This is not the way companies pursue a "rapid application development" practice today. Furthermore, ethical IT and AI design need decentralized, data-scarce, accessible and transparent IT architectures, controlling for system risks and limiting many data-sharing practices. This runs counter to most companies' business models.

Furthermore, when it comes to AI, specific emphasis is laid on the value of transparency. In ethically oriented engineering circles, it is believed that AI should not consist of black boxes where nobody knows what the system is doing. At least for official system auditors, but ideally also for private users, there should be a reliable explanation as to how a system calculates its results. However, when I discussed the opaque nature of the digital fabric above, I showed how difficult it is to create such transparency. Making AI *interpretable* would imply going away from the kind of complex deep neural networks players like OpenAI rely on. Making their deep neural networks or "LLMs" (large language models), therefore, at least *explainable* is an ongoing challenge that has not yet been surmounted. Explainable AI means producing a model of what is happening backstage in the actual LLM. But this model is in itself always only an approximated representation of what's going on. And so true transparency in a literal sense might never be reached if we continue to persist with the neural network technology we currently have.

6.1.1 Which Value List Is the Best One?

One further question when looking at all these current ethical design efforts is: which list of values is actually the right one to follow? If different associations, companies, political institutions and countries all put together their own lists of value priorities, what should IT companies choose to rely on? Behind the scenes there is even competition as to who will be the pioneer in ethical design and who can assert themselves with their list. This is so pervasive that the *Center for the Alignment of AI Alignment Centers* has been created as a mordant parody of the plethora of different AI alignment initiatives.[256]

In order to bring a bit of scientific distance, clarity and logical structure into the many proposals, I have made the attempt to logically integrate existing value lists for digitalization.[257] My goal was to rank frequently cited values, and in doing so I used Maslow's pyramid of needs. Maslow's hierarchy of needs allows us to see which value needs are actually of particular importance for human self-development.[258] In the traditional representations of his pyramid, health and safety are ranked as more fundamental than community and dignity; for example, Maslow observed that people often see no sense in striving for higher values such as prestige when they are ill. Similarly, the politeness of a robot working in a nuclear power plant is of little intrinsic value if the plant itself does not meet the highest safety standards. However, Maslow also saw that the hierarchy of values is not rigid. One cannot say in principle that one value in the pyramid is always more important to a person than another simply because it is more fundamental in the pyramid. Someone may be gravely ill, yet their dignity can mean so much to them that they would rather die than suffer indignity. The relative importance or height of the values must therefore be weighed in each context.

Little attention is also paid to the fact that Maslow's pyramid of needs does not just list any old value, but only *intrinsic* values. These are foundational values that are so important for human development that they are valid across all cultures. Even though cultures practice them with different norms, they are relevant for all people. Interestingly, such higher intrinsic values are usually driven by or enabled by lower values. A high value, such as the security of an IT system, requires a range of lower value qualities that make it possible; for example, secrecy and accuracy of data or responsibility of the operator. A high value such as trust (in an IT system) requires a lower instrumental value such as transparency. A high value like justice requires the lower value of absence from bias. Because of these logical value chains, in addition to Maslow's own entries, my aggregated value pyramid also includes many of the lower value qualities primarily discussed in the technical community, such as transparency or freedom from bias.

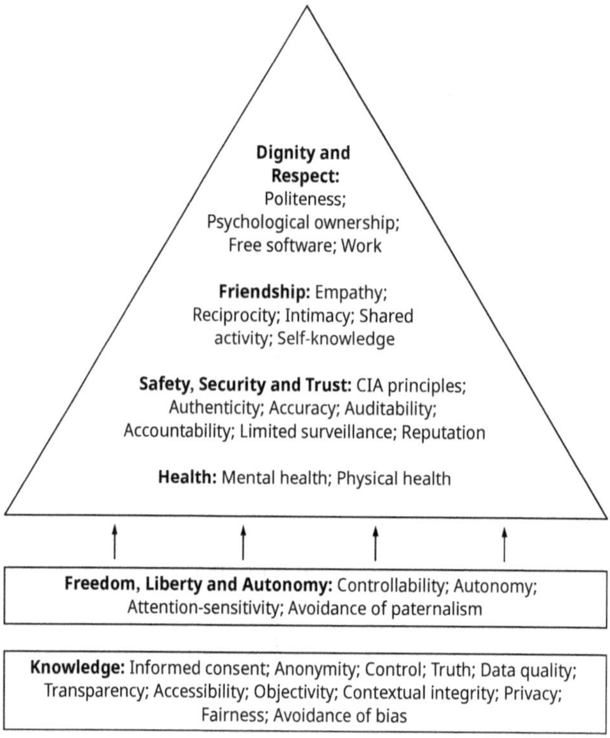

Fig. 6.1: The Maslow pyramid of needs for the digital world.[259]

Is my integrated pyramid of values now a *fait accompli*? Suppose everyone agreed on this model: would that be a solution? I hope that the IT world will agree in the long run to at least consider these high intrinsic values for human self-development as minimum requirements in IT and AI design. The first signs of such developments are slowly emerging on the horizon. For example, the search engines Startpage or DuckDuckGo position themselves as privacy-friendly search alternatives to Google; similarly, the messenger services Signal and Threema differ from others on the market with their discreet and secure communication. Proton, a Swiss company, offers secure and private mail services and document-sharing. There are cloud providers who advertise their service with particular emphasis on security. And there is the Fairphone, a smartphone manufacturer that differentiates itself through the greater sustainability of its components.

But what about AI services specifically? Can we expect that these will be more ethical from the start since they emerged at the same time as so much value-principles work? I'm afraid it's not that simple. One issue is that with the arrival of software agents like ChatGPT we can expect that virtual AI-characters will be accompanying many of us as permanent virtual "butlers." Access via voice is likely to change the entire interaction paradigm of the Web. In this scenario, few people would continue searching personally any more via services like Google, instead telling their software agents to do so. And the big question is whether these personal agents are then trustworthy, private, decentralized, controllable, etc; or whether they are operated by the traditional tech-monopolies who have not proven themselves to be particularly ethics-prone.

Another dimension is equally important to consider: value lists guiding AI organizations through a checklist of pre-specified values are actually not good enough to build really trustworthy AI; not even if my value pyramid was respected. This is because, in truth, values are deeply contextual and every single AI system or IT system requires its own specific value exploration and value design strategy, as is promoted by Value-based Engineering (or ISO/IEC/IEEE 24748–7000). The next section will explain in more detail exactly how this is the case.

6.1.2 Value Principles Are Not Identical to Ethics

In September 2017, a screenshot circulated on Russian websites: it showed how a Russian-speaking speech assistant and its English-speaking counterpart react in completely different ways to the exact same user statement. The user told his English-speaking Alexa, "I feel sad," and the device replied, "I wish I had arms so I could give you a hug." Russian voice assistant Yandex, however, curtly told its Russian users that "no-one said life is about having fun."[260]

I present this example to discuss the value of consolation. All cultures of the world have a conception of consolation. It is instrumental to higher values such as closeness, community, mental health and joy in life, but also to discipline and grit, as the Russian answer shows.[261] Depending on how a chatbot or software agent formulates the answer, users may have different attitudes to life in the long term due to the way their AI reacts to them on a regular basis. For example, when children talk to their AI and learn in solitary situations that life is no picnic, they will develop different virtues than if they are told to seek a warm embrace.

While we might agree that consolation is a relevant value in the voice assistant's system design, it is not recognized in any of the value lists that are currently circulating in the tech world. Hence, no development team on earth that chooses to follow a list of values will have the value of consolation on their radar.

The example shows that predetermined lists of values are always incomplete. They correspond to a kind of solutionist model thinking that can never fully meet the unpredictable ethical diversity of future situations. This is why the three philosophical questions from Chapter 3, which I teach to my students, are so important. The FoodIS case demonstrates this. As you might recall, one question is: what does the concrete technology in question do in the long term to the character of the affected stakeholders? In light of this virtue ethical question, a development team might introduce an AI chatbot for actively dealing with users in different situations and then considering what values and virtues family members or guests might cultivate. Perhaps in this way the value of consolation would not elude them – especially if they gave their imaginations free rein and had not limited their mental freedom by a prior study of value lists. Of course, there is no guarantee that the three ethical questions would have found the value of consolation. We cannot be sure, no matter the means we employ, to always anticipate all ethically relevant situations that digital devices might face. But that does not mean that we should throw in the towel. We humans can only grasp ethical behavior and values by living our lives and continually laying down strata of new experiences on the bedrock of our selves. Technical systems need the same time to learn. Once the value of consolation has emerged as relevant, we can include it in our technical lists of values to deal with. We can leverage the experience gained from using the machines to successively improve them over many years, refining AI answers and completing value requirements backward. For this we could even create worldwide data repositories with value descriptions that all engineers share with each other. The only question is: Will the manufacturers of the systems participate? Will they spend the money to develop valuable mature AI over time?

Certainly, it is not enough simply to decide that a value such as consolation is relevant and to include it in technical requirements lists. If system developers who as computer scientists have usually not studied the humanities see the value of consolation on their system design requirements list, then they may well ask: what is consolation anyway? How can the conditions for this value be created in a chatbot in a logically clear and ethical manner, while also guaranteeing culturally adequate standards for this phenomenon in global markets? To answer this next and deeper question, we need to take a much closer look at the relevant value. What is consolation to different people? We must conceptually break a value down and understand it in depth. We must do what the Value-based Engineering standard IEEE 7000 calls "value conceptualization." How this works is shown in the next paragraphs, turning, though, to a much more famous value now; namely, electronic privacy.

6.2 Conceptualizing Values: The Example of Privacy

When I first started working on digital privacy in 2001, I had a cursory idea of what that value signified. The first step at the time was to go to local libraries to find more information. But there I initially found only a single book and a solitary article that contained "privacy" in the title. A meager yield that stands in sharp contrast to today's AI results (e.g., ChatGPT answers), which immediately provide users huge chunks of information available on any topic.

Besides asking AI systems to explain values, it is also natural for us to understand them in terms of what we have learned about them in the course of our lives. We often argue that we "have" certain values or "hold" them; perhaps because our parents explained them to us. We even think that others must share in our intuitive understanding of them. However, as I outlined earlier it is not fully true that we can "have" values or possess them in any way. In truth, they are abstract invisible ideas that we recognize in the world. We cannot arbitrarily define them according to our own rough ideas or instincts. Values are like colors in the world. They have objective qualities and sometimes even a legal history. How, then, can we view privacy in this regard?

The oldest legal ruling on the subject of privacy in the Western world was passed by an English Court in 1604, stipulating that no landlord or king was allowed to enter one's home without invitation.[262] Until then, the authorities had been allowed to enter private homes without prior warning. This right was abused, and for the ordinary English people it was a great step forward to know that their home would be their private domain in the future; hence the English saying, "my home is my castle." The concept of privacy as *protection against physical invasion* has since been taken for granted in many Western countries (although we should not take it for granted in the face of predictive policing and other similar surveillance practices in development).[263]

In 1890, another conceptual dimension beyond physical invasion was added to the legal understanding of privacy, namely, the "right to be let alone."[264] This was recognized around the same time that cameras became commonplace and many people did not like being photographed by strangers.[265]

The 20th century then saw a wealth of legal disputes surrounding privacy. The history of court decisions has been meticulously documented by the legal scholar Daniel Solove.[266] In his analysis, he has categorized 16 different types of privacy violations in which the plaintiffs' complaints have been upheld in court. Applying some of these examples to the present time, privacy is violated when digital devices permanently surveil people's whereabouts, conversations, actions or other data traces left online; when such data is then aggregated and linked to a person (identification); when data collected is saved in an insecure way or shared with others for unforeseen secondary purposes; and when information about a person is disseminated, thereby disclosing something about a person or even exposing him or her. Any appropriation of this personal data or distortion of a person are considered a privacy breach. And finally, any form of intrusion or decisional interference is considered a legal harm, so constant interruption or nudging through digital services should be on the brink of actually being considered a tort. Taken together, we see how complex a value like privacy is and that we have to have a lot of historical and legal knowledge to act on it both privately and in a technology context.

Against the background of these many conceptualizations of privacy, it becomes clear how superficial claims are which state that "Privacy is dead. Get over it." Such assertions ignore the wide array of disputes on this topic that have been documented and that are taking place. The real challenge is that the averagely educated person today cannot know about all these deeper meanings attached to privacy, or to many other values around. Many believe that ethics is simply about sounding out one's intuitive understanding of values. But this is a misleading way of thinking that does not do justice to the nature of any value. High intrinsic values like privacy are firmly established in the world, regardless of our personal attitudes toward them. We can choose to give them a space in our lives and cultivate them in our family or social groups. Or we can choose to not do so. And we can decide to not give negative values space in our lives. This is the most fundamental definition of ethics: giving adequate space and rank to the right values in our life. But understanding, recognizing, and giving room to values in adequate depth is neither easy, nor without conflict. People often don't agree on value priorities because they have little training on what matters more or less; even though rank orders of values are discussed by scholars like Max Scheler.

6.2.1 Privacy at its Limits: A Current Account

Unfortunately, we are not always in the position to prioritize a value as we deem correct. After the attacks of September 11, voices grew louder that, despite the scale of the attacks, the Internet should not be expanded into a surveillance infrastructure because societal freedom and privacy were more important than protection against a few terrorists. This was certainly a valid argument. But many computer scientists, engineers, and political decision-makers ridiculed the privacy advocates, characterizing them as naive. Since the attacks of September 11, these visions of data-control have indeed been put into practice. Under the surface of societies' visible structures, some major US data management platforms[267] have built up gigantic profile databases of almost all Internet users on earth.[268] Sometimes these data management platforms contain as many as 30,000 entries per person, according to their own reports.[269] Credit card companies, for instance, have been aggregating and sometimes sharing people's spending patterns for a long time.[270] Municipal offices and doctors have resold citizens and patients' data.[271] Internet surfing habits are known by over 170 trackers. And even mails are machine-searched by many providers. So not just abstract data but also all our messages, photos and social media postings are collected and are being analyzed and used by parties we know little about. Indeed, a lot of what we speak about while having our smartphones with us seems to be recorded and shared now, as well as almost everything that we view on video platforms (depending on the service). The permissions for this outrageous secondary use of our private data is given by default if users are based in the US and they would need to proactively opt out if they wanted to prohibit data-collecting and sharing from happening. In Europe the situation isn't much better: here service providers bombard users with endless pages of terms and conditions that no one has time to read, yet which we simply have to sign (opt in) if we want to benefit from a service. If we want to take part in the digital world, we have no other choice than to resign ourself to repeatedly pressing "OK" and opening up a Pandora's box of data-sharing and lost privacy.

Finally, with the arrival of social media and AI an unparalleled intrusion of privacy has entered the arena. Confidentially shared photographs or information is often exposed by peers. Kids expose and blackmail each other, with 1 in 10 minors admitting to nonconsensually sharing someone else's nude photo or video.[272] But avoiding carelessly sharing private photos is no longer enough: with the advent of AI, people can now easily fabricate naked images.[273] (While this may often be intended in a playful or harmless manner, through social media spread and increased accessibility such play easily turns into personal trauma for those exposed.)[274]

In sum, this short account of the history and current reality around the value of privacy shows what it means to "understand a value in depth." If we analyze a value deeply and conceptually dismantle it, a whole world of meaning arises. It becomes obvious why "post-privacy" arguments are superficial. They do not capture the phenomenon in sufficient depth and breadth. It also becomes clearer how important it is for engineering and computer science associations such as IEEE, and indeed all educational institutions, to ensure that young computer scientists as well as existing managers are better informed about value considerations and the legal foundations of their work. Understanding values in depth is crucial because every single dimension of their meaning requires technical measures for protection. For example, maintaining privacy at the technical level not only requires the end-to-end encryption of data to keep information secret, it also requires an automatic and easy exchange of user privacy preferences with data collectors rather than endless terms and conditions. Only by means of simple labels, similar to our food labeling, is it possible for every citizen to know which personal data is collected when, by whom and for what purpose. And in the interest of informational self-determination, it should be much easier for citizens to disagree with companies' data-gathering preferences. This, in turn, requires technical devices and corporate policies which in turn ensure that users' privacy preferences are respected. In addition, simple means must be offered to withdraw or delete personal information; not just on the visible surface of the systems (so someone *believes* the data is deleted) but also on the backend, where backend systems are genuinely deleting it from the databases or from the responsible cloud systems. Such measures are only a small part of what it would mean to truly protect the value of privacy on a technical level.

It should come as no surprise, therefore, that an entire academic discipline with its own conferences, lectures, books and journals has emerged in recent years on the subject of privacy, including the "behavioral economics of privacy," "privacy impact assessments," "privacy by design," etc. The work of this discipline is seeing new challenges as AI becomes commonplace. One such challenge is that with AI systems users no longer configure the privacy preference for each Web service any more in 1:1 relationships between a truster and trustee, but act through an AI agent that then acts on their behalf, and on which they must rely. For example, a user tells her software agent to purchase certain items at Walmart. She has to trust that her software agent (AI) thereby protects her personal data: this ranges from purchases made and preferences to voice data and other personal data. The software agent's operators (i.e., OpenAI, Character.ai, Anthropic, etc.) would then have the duty to not misuse the personal information they receive from people who interact with their new "butlers." But sitting on all this knowledge about their AI users and listening into millions of human-AI conversations must constitute a big temptation for these AI operators; especially in times where such dialogues are considered the "oil" of the new economy and also subject to curious state-driven interests. Incredible knowledge, data and manipulation power is in such close reach that it takes a lot of ethical backbone from AI providers to resist the temptation to abuse it.

6.3 Putting Values into Practice: Resisting Temptation

I have learned the hard way myself how difficult it is to actually act ethically when navigating today's industry realities and not give in to temptation. In the first chapter, I withheld the fact that I did not want to renounce my entrepreneurial life altogether when entering the world of academia. At the very same time as I arrived at the Humboldt University in Berlin for my second book, I was also in the process of founding a small start-up called SkillMap.[275] SkillMap was, as the name suggests, a software that had the ability to map the skills and knowledge of people, whilst also visualizing the networks of who knows whom. You could zoom in on a beautiful graphical interface in a social network and back out again. Initially we just used SkillMap as a coordination platform for our Berlin research center on Internet economics. In 2008, however, came the point when we had to worry about how to make money with this beautiful software. Here I learned firsthand what it means to be tempted to turn away from values such as privacy.

6.3.1 The Problem of Temptation

Technical development, sales and marketing needed to be financed. It was of paramount importance to hire (and pay the salary of) a good manager who could take over the day-to-day operations of the company. So, together with my developer team, I racked my brains over where the money might come from. And, of course, the idea came up to market the personal data collected in SkillMap. We anticipated that the information about who knows whom in a company, who knows what about products and services – or even, who does *not* know what – is a valuable insight for companies, HR managers and executives. Likewise, the information regarding who knows whom better or worse and communicates more or less with others is vital to understand the social networks in one's company. So, to put it bluntly, it would have been possible to measure who is important in a company and who is not; perhaps even far beyond the way home-office surveillance offers it today. After being asked by some potential investors, our product roadmap then included the idea that we could more or less secretly tap into the e-mail traffic of employees in SkillMap, using the software to visualize the lived relationships between users of the tool. This would have meant that the software skills that I devised as an active privacy researcher for my own business were, at least in part, exactly the opposite of what I preached in theory. My deep knowledge of the value of privacy did not help me at all. Preach water and drink wine. The lofty ideals had collided with the harsh realities of the business world.

Luckily, I didn't have to make that decision: we never got any money and the start-up was dissolved. However, I learned from this experience how difficult it is to be ethical in practice. I also learned how quickly we can suppress what is actually ethically correct; in this case, protecting the privacy of users and not giving management too much power. In light of the desire to succeed we quickly dismiss our bad conscience regarding the values we undermine. The temptation feels similar to the desire for a large piece of cake, which, if we're health-conscious, we really ought to turn down. So, I do have real empathy with start-ups and businesses when they face the same dilemma as I did and feel they have to do things that are ethically borderline for their money. Those in the social media industry say they need to make their users addicted so they stick to the platform and look at ads. Digital game producers say their games have to be addictive so players come back and the multiplayer online worlds have enough players. Many managers believe that they have to automate their organizational processes because otherwise these will be too expensive compared to the competition. At the crossroads of true humanistic progress, we face the dilemma that we are externally forced to do things that effectively drive society into social regress simply because our business models are the way they are.

6.3.2 The Will for the Good

This feeling of being under pressure is not a good place to start on our journey to a better future. It gives the impression that there is only one way to be a good entrepreneur. That, of course, is never true. An old proverb says, "Many roads lead to Rome." And so start-uppers, entrepreneurs and investors do not necessarily have to take the one way that makes their business ethically questionable when they participate in personal data markets, or what Shoshana Zuboff calls "Surveillance Capitalism." Instead, entrepreneurs in the digital economy should seek out business models that allow them to adopt more humane models of progress. Facebook could simply charge its users $3 per month, offer a wonderful social networking service and forgo any use of personal data to build and operate addictive models.[276] Car manufacturers could prioritize research funding for energy-efficient vehicles rather than building on AI for self-driving cars. I also think that the history of FoodIS shows very nicely how the company could ethically increase revenue and save cost in a socially beneficial way. On the revenue side, FoodIS could offer food consultation. The company could expand the courier app to a sports app for bikers and market this innovation separately with partners. It could integrate community features and save new recruiting costs as well as attorney fees for litigation as the loyalty and enjoyment of couriers increases. If FoodIS took this ethical route instead of squeezing out the last bit of margin with a narrow-gauge service from the drivers and restaurants then the company would have a different spirit than it does today. And that spirit would surely be common to all employees and market participants, perhaps being such a boon that the long-term existence of the company would be more secure than it effectively is today. As I said, the good is boring and naive only in the imagination; in concrete reality, it is always new and exciting.

If you talk to entrepreneurs from Silicon Valley or anywhere else in the start-up strongholds (Berlin, London, Shanghai, etc.), many have the will to do good. This is much the same as how I was thrilled with the possibility of employees sharing their knowledge in a better way when using SkillMap. Creating "knowledge" was my first value focus. That alone was worth the effort to found the company. Many entrepreneurs, like me, are concerned with developing such primary values for their service. Later, however, their core values are displaced all too easily by pressures for efficiency in implementation. They push people into ungodly workflows, handle personal data in an irresponsible way and support surveillance infrastructures. They then try to play down or sugarcoat all the human and social misery they incur, not because they are bad people or deliberately want to harm this world, but because they often lack the know-how in their mission of how to orient themselves toward a better business model. The latter requires understanding how to professionally develop a truly meaningful value proposition. It requires carefully weighing the long-term human advantages and disadvantages of a new service. In short, entrepreneurs and managers would need to think about the three questions that philosophy can offer to help them, and which I presented in Chapter 2. At the moment, they have no idea on how to harness the power of values and engage in Value-based Engineering. Had I known this back when we developed SkillMap, we would certainly have arrived at a friendlier business model and technology design.

On top of all this, however, there is yet another real ethical challenge to goodwill: the "weighing" of values. For true human progress, we have to weigh values. I showed this in the second chapter when discussing the Disney laundry, where Disney has to weigh profits and efficiency against values such as dignity and privacy. Today, it is often argued that for the realization of one value one must necessarily sacrifice another value. For example, it is said that patients will have to sacrifice privacy in the service of enabling health research; that patients would have to share their health data, otherwise no progress will be possible. Or it is often claimed that state security can only be guaranteed if the government has as much of its citizens' data as possible, so that it can calculate and predict crime. It seems as if security is only possible at the expense of privacy. Few realize that the world of values is not a world of trade-offs and mutual exclusion, but a world of balance. One value does not exclude another value. Instead, being a good and wise world designer is about creating the right balance between values. If you go to extremes and only create one value at the expense of another, then the progress curve in Fig. 1.2 cannot be realized. It leads, at best, to directionless change – perhaps, to "travelling on the spot," to use the words of Paul Virilio.[277] For the curve to move upwards, we must learn an ancient art that we have forgotten in our time of abundance and extremes: the art of omission.

6.4 Value Creation and the Art of Omission

The art of omission is a real skill. Bringing values into the right balance through omission requires the sensitivity to recognize negative values and to omit them. How difficult that can prove today is evident in the field of healthcare, which could be truly revolutionized by digital technologies.[278] Digitalization allows us to collect an infinite amount of data about the human body, about healthy behavior, physical activity, emotional reactions, stress, genetic engineering and much, much more. Building on this, digital analysis tools, often powered by AI, seem to allow us to understand many connections between our biological nature and different life situations. My dream would be to one day use our digital knowledge to combine the holistic insights of Indian Ayurvedic practices and traditional Chinese medicine (TCM) with Western medicine. Other visionaries want a completely individualized medicine with personalized medications. The positive potentials of AI analysis for the value of health are so promising that I personally would be willing to share my health data regularly for such a good cause. The value of health could contribute to a steep positive progress curve in our society. However, this positive medical future is realistic only under political, technical and economic conditions of omission that are wisely chosen.

6.4.1 Forgoing Half-Truths and Secrecies

First of all, the sharing of health data would have to be done with the explicit consent of the data donor(s) and not secretly or under duress. Many physicians in German-speaking countries have been selling their patients' marginally pseudonymized medical records to large data brokers in the healthcare sector for years.[279] What is even more questionable is that Google's Deep Mind has simply sourced the health and care data of 1.6 million UK patients registered with the National Health Service (NHS); patients were not aware that their health data had been shared, nor had they agreed to participate in any research.[280] When things like this happen, it only fuels negative responses such as mistrust and helplessness, accompanied by feelings of anger and frustration over a loss of control. In the fog of opaque half-truths about data collection operations, all those involved fall into autocomplete mode and imagine what nasty things Google could do with the data. Could insurance companies get access to the data as well and even adjust premiums? Could intimate diseases become known which we have always wished to keep secret? Many thoughts can arise. If privacy is secretly sacrificed at the altar of health research, as in this case, there is no ethically welcome progress. Negatives are superimposed upon the positive. There is no sustainable cooperation that allows for regular research and shared knowledge for better health. This is a sad story and one that should not be repeated in truly progressive research environments. So what must be done? What is the right conduct?

The ISO 26000 standard entitled "Social Responsibility Guide" asks companies the following question: would you be comfortable if your behavior became public knowledge?[281] Google and the NHS would surely have had the opportunity – in the sense of the art of omission – to forgo any of the secrecy. They could have used a transparent way to get prior consent from patients seeking to participate in their health research. Technical models on how to organize data-sharing for research purposes have been around for a long time.[282] People consent to their data being used, they can withdraw their consent if needed and typically, they receive some kind of compensation for participation. Such fair data exchange can then, of course, be made public at any time. There is nothing to hide. In contrast, the NHS shared patients' data in secret and the practice was then uncovered only after investigative journalists brought the truth to light and laboriously pieced together what data was shared.[283] Against the background of this example, the first lesson for the art of omission is to *leave out stealth.*

6.4.2 Forgoing Structures of Temptation

A second aspect of omission appears in the same Deep Mind/NHS case: it involves the effort, time and money that would have been required to share patient data openly and transparently. To do things in an above-board manner, it would have been necessary to establish a transparent technology infrastructure between the NHS, Google and the patients in order to organize the data exchange. But such an infrastructure would require that Google and the NHS not only make the necessary IT investments but also respect that some patients may not want to participate in the research. Likewise, they would have to respect when patients wish to provide their health data for only a part of the research or wish to limit it. If such privacy preferences for the data subjects exist, the recipient company must resist the temptation to store more data than was granted, or to use it for purposes other than agreed.

In practice, this temptation can be so great that it has spurred the European Data Protection Regulation in May 2018 to order privacy by design for many digital systems. This means that technical architectures and routines in an IT system (including AI) should be designed from the outset in such a way that it is no longer possible to do much more with the data than is consented to by users. In Germany, such a technical practice of limiting data use has existed for a long time: telecommunications providers must delete citizens' call data records once the billing purpose is fulfilled.[284] What is no longer there can no longer be analyzed. As a result temptation is avoided. The second lesson for the art of omission is therefore to omit IT architectures that could undermine values; or vice versa: to build IT architectures that help us to control ourselves. Regular data deletion routines are an example. Such ethical strategies, as I recommend for IT here, we all know from everyday practice as well: If we want to lose weight in the interest of our health, we clear the fridge and avoid buying anything that could tempt us. We build an environment that helps us to avoid temptation.

6.4.3 Forgoing Excessive Profit

The third condition for progress in the field of health is that it benefits the whole of society. As I will show, nowadays a lot of health knowledge disappears in the walled gardens of companies' "trade secrets." Often, only very wealthy people can afford access to the expensive drugs that are brought to market. Take, for example, a disease called Spinal Muscular Atrophy, which affects between 1 in 15,000 babies born today. The most effective cure is the drug Zolgensma. However, one course of treatment is listed at $2.1 million, which puts it out of the reach of most parents or healthcare providers.[285] The example brings an important value to the fore, a value which is certainly placed on top of every pharmaceutical company's list: the common good. But it also brings a vice to bear that must be tackled: greed.

The challenge put forth here is that companies are subject to a one-sided profit-maximization logic: the digital systems and algorithms that today make up an estimated 90% of global stock market trading[286] ignore values such as solidarity, compassion, justice and the common good. As a result, companies are putting a lot of weight on the value of financial profit, which in turn can lead to such eye-popping prices. Profit is fundamental from a business point of view. It is an indicator of a company's health. It also allows managers and entrepreneurs who carry the risk and responsibility for the success of the company to provide wealth. But at the same time, it should also be balanced. The second chapter has already shown that profit maximization as such is a low value. Maximizing it alone will not bring us true progress if it undermines other higher values, such as the common good.

To come back to the case of Zolgensma: the value of profit is here in conflict with the values of the common good, solidarity and justice. The managers of the company therefore have to ask themselves how much profit they are willing to give up in favor of a more positive social value balance. According to Aristotle, someone is generous when he gives in relation to what he owns. That is, someone should not be wasteful on one side or renounce profits to an excessive extent on the other. Medication does not have to be given away for free. But entrepreneurs and shareholders should also not be too greedy either. Medicines that cost $2.1 million lead to social indignation, hatred of the elites, feelings of injustice and despair over unnecessary paralysis or the death of people who could have been saved. As a whole, fewer people will be healthy than might have been possible, and the bigger picture is that all this leads to a social decline. It is not society that benefits from the progress of health research in such a guise, but only a select few. So the third omission lesson is the art of omitting excessive profit.

6.4.4 Forgoing Premature Products and Services

Digitalization and AI technology bring with them the promise not only of revolutionizing our knowledge and health research but also inventing new individualized medicines and working as a community for a healthier future. Furthermore, many new devices are being brought to market, along with mobile health apps. We can buy and use these to improve our day-to-day life, or so we are promised, at least. An example of such a device is a cap or headband used for so-called "transcranial direct-current stimulation" (tDCS) of the brain. The device applies electrodes through which a current flows to stimulate brain activity. The goal is to activate various regions of the brain in order to improve memory and foster concentration. Additionally, manufacturers promise, depression and chronic pain will be alleviated. That sounds fantastic; clinical research in the area is ongoing. But, unfortunately, research results are ambiguous. In July 2019, 39 leading neuroscientists warned in the *Annals of Neurology* that the use of these caps should be approached more cautiously; after all, little is known about the long-term effects of such stimulation on brain functions, and negative neurological consequences resulting from the use of such caps cannot be ruled out.[287] Despite these concerns, anyone can buy the devices on Amazon for $159.97. This is ethically problematic.

Companies that bring technologically immature products to market lack patience. They want to sell a technology they only just conceived of, to work right away on the market. They might be pushed to such a precipitous decision by the wish to meet their investors' aggressive schedules. And this is true whether we're talking about transcranial devices, drones that fall on people's heads or self-driving cars that kill cyclists. If software development follows the maxim "release early, release often," then many in the IT industry seem to believe that this can be transferred to any tech products. They underestimate the time it takes to test products, carefully monitor their effects, and, if necessary, wait many years for their proven maturity so that they can be brought to market with good conscience. In contrast to the pharmaceutical and food industry, like a restless child, IT still has difficulties with being patient. It still has to learn the art of shunning the unripe fruit.

6.4.5 The Role of the State and Citizens in the Art of Omission

Practicing the art of omission is an ambitious goal in today's digital economy. Is it therefore at all realistic to expect that companies will advance digitalization in a more careful and ethical sense? I doubt it. Many companies have become accustomed not to think proactively in an ethical and value-oriented way but to get guidelines from legislators for ethical action. All over the world, ethics is outsourced to the legal regime (if there is one).

Legislators are therefore required to protect the preservation of high values. They must ensure that not only values such as privacy and health but also knowledge, freedom, community and dignity do not come under threat in the digital age. They are expected to set value priorities and have already done so in some regions of the world – the European Privacy Regulation being one example. But government could go further and support the four arts of omission outlined above. They could impose strict transparency obligations on companies. They could stipulate ethics by design in IT architectures, including AI services. Furthermore, they could meaningfully tax and redistribute corporate profits, and impose stringent licensing rules or set price-caps for products and services fostering the common good.

However, it would be wrong to think that only governments can smooth the path of values for us customers and citizens. For too long, we have faithfully delegated our progress as individuals to these external forces, relying on their good intentions and protection. However, we live in a time of brutal capital markets; there is tough global competition and companies as well as state authorities find it harder than ever to practice the art of omission. We also have to realize that, given their financial power, more and more companies are not shying away from simply breaking laws or operating from jurisdictions where there are no sanctions. Therefore, if we as citizens do not take small daily steps to consciously live out our values by taking corresponding buying decisions, then we cannot really hope for progress. Human progress in the age of AI will be, to a large extent, also about how we as individuals in our different everyday roles (as individuals, clients, employees or entrepreneurs) act and use technology. Before dealing with this field of personal engagement, however, I would like to discuss two values whose development is particularly important for progress in the digital age: knowledge and freedom. Without them, as Maslow has already shown with his pyramid of needs, there can be no self-development or human-centered progress whatsoever.

7 Knowledge in the Digital Age

We attach great importance to theoretical models of nature that enable us to advance scientific and technical progress. However, we often forget that knowledge is to a much larger part an expression of practical doing. Knowing how to raise a child, bake bread and arrange a garden is just as relevant a knowledge as the specific knowledge of a Nobel laureate.[288] Both forms of knowledge, model-driven theories and practical life knowledge, strengthen the self and give us a positive awareness of who we are.

On the Importance of Human Knowledge

Knowledge is not only important because it empowers us in our endeavors and gives us pleasure, it is also important because it is what we remember and know that determines who we are; that makes our identity. Proper knowledge is not the same as our ability to consult Wikipedia. Although Wikipedia is a sensationally practical technology, without which my professional life would surely suffer, it does not replace the need for knowing things myself. I often have this discussion with students who do not understand why memorizing exam material is so crucial. Many believe that such learning is not important – why spend all this time struggling with books when they can access Wikipedia or consult ChatGPT? In contrast to AI, however, our human memory does not consist merely of facts. Instead, our long-term knowledge consists of a mix of value-laden memories that we retrieve in various ways. These then make up our personality: memories of select facts, of important events, of appreciated stories and of impressive people. Our long-term knowledge is gained more through emotional experience than it is acquired through rote memorizing.[289] If we understand knowledge formation like this, then it is clear that not everyone remembers the same lifeworlds, including their factual elements. In fact, we all finally sketch something different from what is taught and written, and in doing so, we become unique persons.

So, if we think it is not important to train our own long-term memory because we find everything we need through personal AI agents, then this argument is comparable to arguing that we do not have to play soccer anymore if we can watch it on TV. People who think and act in this way will probably no longer have a sufficiently broad long-term memory as they grow older. They will know less and as a result, they will literally lose out on who they could be. As I described in Chapter 3 when discussing the "Borg fate," these people's personalities will literally shrink because they have less knowledge, autonomy and self-confidence to stand up for themselves.

 | https://doi.org/10.1515/9783111677736-007

Expertise as a Precondition for Participation

In addition to joy and identity, there is a third reason why human-based memory is so important: the social dimension of knowledge. Knowledge determines our place and role in a community. If I train my students in business informatics then it is with the assumption that they will eventually belong to that professional group. Each discipline has a shared language, with its own shibboleths.[290] This specific vocabulary shapes the identity of a discipline or craft. It is therefore impossible to want to belong to a domain without having internalized the corresponding ways of thinking, speaking and writing in technical terms. Even though we live in a time of increasing interdisciplinarity, it is presumptuous to think that I could just have a say in a subject that I have not studied for years. What I mean by *studying* is not necessarily university education but rather the period of repetition and exercise that I have to invest in a community with others in order to internalize the ideas, terms and methods of a profession. Only the disciplined appropriation of a subject-specific "craft of thought" allows me to "go beyond" what has already been achieved.

If we put all this together, we realize that it is our own individual knowledge that makes us happy, shapes our communities and gives us the opportunity to participate in society. Subsequently, it becomes clear that little is more important for true progress than to develop technology (and specifically AI!) in such a way that it helps people build as much of their own knowledge as possible.

7.1 How AI and Digital Services Can Support Humans' Knowledge

Above all, AI is promising that it will increase our theoretical knowledge of the world. Massive amounts of data can now be collected on life at the micro and macro level. For example, we can discover in unprecedented detail via sensors and modern optics, the very core of the body – patterns that occur in biology on finely granular levels. At the macro level, we can capture highly complex social dynamics that would earlier have slipped under our radar.[291] This data, coupled with AI technology's pattern recognition power, can be used to form new hypotheses about the world that we can then scientifically investigate. In pursuing this (AI-supported) powerful path of discovery, existing models of the world are sometimes turned upside down. All this is exciting and positive. Many speak of a knowledge explosion. While 250,000 English-language books were published in the USA in 2002, there were over 3.2 million in 2022.[292] The number of patents is also increasing. Currently, more than 3 million applications are made worldwide every year.[293] Many innovations in the field of individualized medicine, materials research, global networking, transportation and nutrition stand as positive consequences of this development.

The explosion of publications is not, however, a purely positive phenomenon. Publishers have been inundated by "authors" who are now using ChatGPT in order to make a quick buck rather than having a genuine interest in authorship or the value of literature.[294] In the academic world, a plague of fabricated papers from "paper mills" is emerging with the help of AI. Paper mills sell authorship on papers that often rely on AI chatbots to produce text.[295] A report published in *Nature* estimated that 400,000 research articles published in the last 20 years may have been sham studies of this kind, bought to boost publication records.[296]

That said, if we presume that there is still an overall increase in knowledge, then sharing this knowledge is also a huge opportunity. I often think of one of my acquaintances, who for a long time was responsible for the Business Information Center in Uganda at the United Nations.[297] She told me about the tremendously positive influence that Wikipedia has had in Africa. For a long time, the region for which she was responsible had been plagued by a plant fungus that had decimated their crops. By providing Wikipedia access in the UN regional IT centers, the rural population learned that the plant fungus could be controlled with a simple aeration process. And, as people applied this learning, the problem disappeared.

I could imagine that even if people in developing countries did not have smartphones or computers to access AI in the future, they might still be able to access AI services and get advice on matters that burden them, like fungi destroying their harvest. About 84% of adults in low- and middle-income countries in Africa now own cell phones.[298] Hence, AI advice could be at these people's fingertips and tremendously support them with a level of advice they might have never had before. But, that said, one must also be cautious, because AI systems do make mistakes, as has been outlined above.[299]

Easy access to knowledge is not only fruitful in developing countries – it's a blessing to me too, living in an industrialized and rich country like Austria. At the Vienna University of Economics and Business, I have direct access to 20,000 scientific journals, which I can find within minutes and then quickly download. When I wrote my diploma thesis in the 1990s, I was only able to access journals via extremely unwieldy microfilm sheets. I could gather only a fraction of the existing knowledge with months of laborious work and moreover, I was never sure whether there might be relevant works that I missed. Reading, looking up, understanding and learning was very tedious. Even though they are private companies, and not strictly "open source," platforms such as Google Scholar, Google Books, Academia.edu, as well as learning platforms such as Coursera, the Khan Academy, edX or TED Talks are a blessing to my daily work. And sometimes these days I even integrate AI chats, asking ChatGPT or other services for overviews and summaries of issues that I am not an expert on but need to understand quickly.

To get back to the picture of true progress: knowledge distribution is receiving a tremendously positive kick through AI and digitalization. But AI and digitalization in general are also characterized by an overabundance that can tip the value of knowledge into a negative spiral. In fact, we might stand more at the crossroads than most would expect: we have much more knowledge at the societal level than ever before and yet, at the same time, on an individual level we know less than ever before.

7.2 Why We Are at the Crossroads of Knowing

In order to create true progress in this digitalized AI world, we must take a sober look at reality, despite all the positive developments. There are a number of problems regarding knowledge formation that could hinder true progress in the future. These include unnecessary restrictions on access to knowledge, a knowledge crisis in research, the rising influence of popular science, and the increasing inability of experts to actually use machine knowledge fruitfully.

7.2.1 The Political Problem of Knowledge Access

Many authors have brought up the subject of a "digital divide" in the past, arguing how important it is for all regions and people to have access to the Internet. At least as important is the question of what we can find online in terms of knowledge. As of 2024, only 40% of the scientific articles published by academics were open access.[300] And while this number is more than three times greater than in 2014 (where only 14% of publications were open access), it still means that over half of the world's scientific knowledge remains behind gigantic paywalls. Large publishing groups such as Elsevier provide access to their online archives only at extremely high prices: a package of journals can cost as much as $28,000. Even many universities in rich industrial nations cannot afford such sums.[301] What is more, in order for an author to have his or her article made open access, he or she as to pay a fee of over €1,500 to the publishing house, though the knowledge is submitted free of charge to these publishers. Typically it has also been proofread by the authors, largely typeset by them and reviewed by other volunteer academics. This whole publishing system works only because many scientists are still funded by governments, who use taxpayers' money to finance the profit margins of publishing houses. In other words, academic publishing is a good business to be in: it privatizes public property by restricting access.

There is also a problematic development in the funding of research. While many still believe that research is mainly conducted at public universities, over 60% of the research is now privately funded and a lot of the most relevant ensuing knowledge is available only to the companies that develop it.[302] Companies like Google or Apple put huge sums of money into in-house research. That's good in principle. They buy the best scientists from the world's best universities to do research for them. And when it comes to AI knowledge, a lot of studies are being published by leading AI providers like OpenAI and Anthropic. "Open source" LLMs have been shared, i.e., by Meta and also OpenAI.[303] The problem is that, at the very least, this sharing of insights represents only a fraction of the knowledge that is actually available. Data analysis and AI training still happens for the largest part behind the soundproofed walls of nondisclosure agreements, inaccessible to the public. As a result, a lot of what is currently being learned by companies around the world is not shared with academia or the public at large. Instead it stays unreachable, and invariably wrapped in a vast shroud called "corporate secret."

This is surely a tremendous loss in terms of social progress. How much could society gain if academic institutions had access to the data collected by big tech? For example, insights into the complex psychological effects of social media use could be gained. Value-alignment issues between AI and users could be studied by the whole world instead of a few specialized research departments at the big companies. Imagine, also, how positive a systematic and institutionalized cooperation between the major players in the IT industry and universities at the teaching level could be in improving the current knowledge of students!

If we look at these developments in knowledge criteria and access, we cannot help but notice a political failure. In Europe, governments are driven like cows through the village when it comes to creating expensive broadband access and funding universities that train top graduates. The knowledge developed by this infrastructure is then privatized and not returned; except, of course, in products (such as expensive medical services) that often only high earners can afford. While the public sector finances almost all of the productive forces necessary for progress and heavily invests in the value of knowledge, the monetary advantage is then skimmed off by a few players in industry. Given these developments, we cannot rule out that the most advanced AI systems in the future that are based on expensive knowledge will also become very exclusive and barely affordable tools that will only point users to sources they cannot afford to access.

7.2.2 A Knowledge Crisis in Research: The Problem of Nonsense

Assuming that we were able to improve the access to knowledge and make more of it available to the many who need it, there is still another problem that hinders true progress: that is the problem of nonsense.[304] Nonsense is related to three phenomena:

1. The Big Data Illusion
2. Senseless questions
3. The rise of overconfident experts, half-knowledge, hubris and neo-tribalism

The Big Data Illusion was described in Chapter 3: there is a danger that what is proclaimed as "knowledge" is, in fact, not. This problem occurs when the data quality of the IT systems used is not good enough. Equally, software mistakes and errors are frequent. As a result, research results can sometimes simply be wrong, leading to nonsense theories or explanations, which may be even more damaging, creating more confusion than elucidation. We may be able to improve on this if we become more aware of the phenomenon and act accordingly; for instance, by putting more time, money and care into data quality, research assumptions, conclusive arguments, etc. But there are two other (quite different) challenges to true knowledge progress, namely, senseless questions and expert overconfidence (hubris).

7.2.3 Senseless Questions

Ten times the number of books and essays published does not guarantee that wisdom will be found in any of them. At the macroeconomic level, it has been observed for years that the productivity of science is falling across industries.[305] In other words, what the average scientist is substantially adding to our knowledge base is on the wane. There is a great deal of writing, but there is a common belief that the value of these copious contributions is small. Take the concrete example of chip technology. Progress in this central area of IT has been described for decades by Moore's Law. Moore's Law states that research and development doubles the performance of computer chips every 18 months. However, today, we need 18 times as many researchers to uphold this rule of progress as we did in the 1970s.[306]

Digitalization simplifies the collection of data, the analysis process and the publishing process in creating knowledge; however, it does not replace either the thought process or the art of asking the right questions. On the contrary, the nature of the digital causes scientists to struggle with shadowboxing and constant interruption, undermining their ability to focus their thinking. The mass of publications made possible by IT can hardly be mastered. Who is supposed to read more than 3 million new books in a year? And above all, IT is changing how people read and process knowledge. Studies have found that reading comprehension is better when reading physical rather than digital media. And the more people read online the less they comprehend, a trend that has increased progressively between 2000 and 2017.[307]

In 2008, the British Library commissioned a study that sought to trace how scientists (professors, lecturers and doctors) read. This was with a specific focus on scientific articles. It found that as early as 2008 (before social media and AI were prevalent), about 60% of the studied scientists did not look at more than three pages from online (digitally displayed) journal texts, and 65% did not come back to a text to scrutinize it more intensively. The report concluded that it is clear that people do not read online in the traditional sense. “It almost seems like they go online to avoid reading . . .”[308] the study adjudged. Since this study was published, a so-called “shallowing hypothesis” has even been put forth, which claims a general reduction in reflective thinking as a result of social media and messenger use: a trend aggravated by AI, which now offers quick summaries of any book or source.[309] Around a quarter of AI users say that they never or rarely click the source links to check the information they receive through a chat.[310] And what is worse, the loss of this reading and scrutinizing lead to a negative impact on cognitive ability.[311]

Given all these developments, if we do not rebuild and maintain a culture of attention around our knowledge we will eventually float like disoriented fish in a sea of information, the quality of which we can no longer judge. Our personalities will be too shallow to ask any good questions because we did not take the time to build up the necessary memory apparatus. As claimed above, there could be a huge gap arising between those few humans, on one side, who really are self-motivated to engage with knowledge and information, continuously developing themselves, and engaging with AI on a level-playing field; and on the other side, the vast majority of humans who will sink back into immaturity (German: *Unmündigikeit*), not willing to read and think any more, using their AI agent for every thought they need to form, replacing their own cognitive effort. The gap between such a meritocratic elite that defines itself through its intellectual ability and knowledge, on the one side, and the vast majority of lazy AI users (“Borgs”) on the other side, perceived as stupid pets by the elites, is already in the making. The consequences for democracy are devastating.

7.3 Overconfident Experts, Half-Knowledge, Hubris and Neo-Tribalism

To cultivate one’s own knowledge, one needs “muses and cult,” as Josef Pieper has argued in his book *Leisure: the Basis of Culture*. For Pieper, leisure time means, on the one hand, to delve deeply into texts, think, reflect and allow for the muses to inspire; while on the other, cult as to when and where to engage in one's own education. Every day, I visit my own inner *scholē* in the early morning hours when over a hot coffee in bed I spend an hour or two reading good books or scientific articles that I need to know about for my work. Scholē (σχολή) is the ancient Greek word for "school" and literally stands for leisure or muses. The cult may be in the coffee, without which I would not feel motivated.

Recently though, as I look into scientific articles I find myself coming across more and more shallow thinking or even nonsense. I observe that we are increasingly suffering from shallow, fake science, which is being brought into the world by preeminent "experts" and poorly-educated "scientists." As an example, I would like to share the story of Anna,[312] who graduated in 2013 with a bachelor's degree in psychology from the University of Michigan. At the time of writing her online scientific contribution, Anna was doing her masters in Cognitive Neuroscience, and as things stand, her university had a working partnership with one of my favorite online magazines, *Aeon*. At *Aeon*, scientists, experts and thinkers can share their insights on a wide variety of science and humanities topics, and I have often benefited from their excellent online magazine.[313] A testimonial was placed next to the donation button on *Aeon's* Website, which stated that the magazine stands for exactly the values the Internet seeks in its purest form: the selfless distribution of wisdom.[314] And now along comes Anna, who is taking her first baby steps toward research in writing her master's thesis and is already allowed to publish an article, which is entitled "How brain stimulation can boost our memory when paired with learning." Sounds great! The title's euphoria is bundled in with an optimistic outlook on transcranial direct stimulation of the brain. Anna thinks that learning to use the attached electrode cap might soon make us much smarter. She references a few studies that bolster her claim with facts. Among sci-fi fans and in transhumanist communities, such caps are of course all the rage. So, Anna is part of a fan community sailing under the flag of science that makes sure that our budding academic could publish her wise essay on *Aeon* before she had even completed her master's thesis (as it seems, at least from CV information) and learned the craft of science. The problem, though, is that roughly around the same time of her essay, 39 leading neuroscientists called for a slightly more cautious approach to the caps in the *Annals of Neurology*, a top-ranked journal of Neurology since the 1990s (see also Section 6.4.4). After all, they argue, we still know very little about the long-term effects of these stimulations on brain function, and we cannot yet rule out long-term damage resulting from use.[315]

The Anna example should be read as a general symptom description for self-proclaimed expertise rather than as an isolated case targeting any specific individual: the magazine *IEEE Spectrum*, for example, read by roughly 480,000 IEEE members, recently published a similarly euphoric article on direct stimulation caps, with limited (but at least some) mention of ethical or health concerns.[316] The examples show that we have a general quality problem at various levels of scientific publishing, whether online (*Aeon*) or in edited print media (*IEEE Spectrum*), whether with presumptive students or already established scientific journalists. The quality problem is not only due to a lack of knowledge but also to a lack of care coupled with a fine dose of overenthusiasm for technology deeply rooted in our Zeitgeist. Semi-scientific experts portray science (especially on YouTube) in a popular fashion, as if truth were so easy to grasp. Science is portrayed as if data evidence always pointed clearly in one direction, as if enthusiasm were always perfectly placed. Thereby, a lot of misjudgement, premature enthusiasm and half-true stories about technical progress are manifesting and spreading that then "inform" others, raise fantasies in the wider public, and lead to false investment decisions and even laughable political decisions. The latter is specifically observable when it comes to AI's superintelligence powers. Fueled by AI tech-leaders' marketing slogans around the imminent arrival of Artificial General Intelligence, many (even intelligent) people start believing in cold silicon springing to life somehow soon.[317] The data centers, they seem to believe, will literally be waking up on them. In a similar vein, the government of Albania recently even appointed an AI chatbot as minister for public contracts. The bot's name is "Diella" and stands for "sunshine."[318] But the fact is that dead materials don't simply spring to life. And I would claim that AI will continue to have insurmountable value-alignment problems because it cannot realize relevance and has no access to meaning: such views are, however, simply shoveled aside by the half-experts who draw their knowledge from YouTube videos and tech companies' marketing shows. The French philosopher Bernard Stiegler saw clearly what we are experiencing here: a "proletarianization of knowledge."[319]

One driver of this proletarianization has actually little to do with poor sources, premature authorship or knowledge creation in the first place, but rather social roots: since the arrival of the first-generation AI in our social media feeds, people are artificially joined together in tribe-like communities that share preferences, scientific interpretations and ideological views.[320] Many groups today, such as the Transhumanists, even build their own online news portals, hold their own conferences and venues, publish their own books, and find enough supporters online to feel empowered to market their vague speculation as community knowledge. On their beautifully crafted websites, soon to be found as AI services, everything looks professional and serious. Bot farms are hired to promote views, boost hits to sites and spread the word. The veneer captivates, but the content is meritless. Without deep knowledge, a difference of opinion is hardly possible. Journalists who are not experts in their respective fields must feel attracted to certain tribes. They follow their preferred influencers on YouTube or they promote what their gut feeling inclines them to or where the IT marketing machinery is currently funding headlines. So, even if funded by public money and working for public channels, journalists accidentally help a great deal of academic nonsense to become somewhat mainstream knowledge. This is a development that is disastrous, and will inevitably lead us into a huge downward spiral of public knowledge. We no longer know but are terribly confused.

7.4 What Is Knowledge?

The confusion surrounding what we call knowledge shows that we need to delve deeper into the term. In antiquity, Plato once defined the theoretical part of our knowledge as "a justified, true belief" in something. All three components of this definition – justification, truth and belief – have been the subject of heated philosophical debates over the past two-and-a-half thousand years. What is enough to justify knowledge? When is something true? And is conviction the same as belief in something?[321]

One of the founding fathers of psychology, Wilhelm Wundt (1832–1920), developed a conception of knowledge that can be useful to us in the digital context. He distinguished four stages of cognizance: faith, opinions, presumption and knowledge. At the first level, we are talking about a subjective holding-to-be-true; a purely personal faith in something. At the second stage, Wundt places a relatively more objective holding-to-be-true based on facts, which he – interestingly enough – called "opinion." Facts or observations of the kind we can learn today from Ted-Talks, good journalism or even in academic papers is hence placed by Wundt on this level only. If such opinions refer to the future and predictions are made on their basis, then this becomes a "presumption" for him: the third stage. So if we predict that based on current AI-supported breast cancer diagnosis the disease can be defeated, then this is a presumption. Wundt argued that such a presumption, however, is still not yet "knowledge" (the fourth stage).[322] So, he effectively denies the status of "knowledge" to a scientific paper, which is merely based on current insights. Scientific opinions or presumptions only enter the realm of knowledge when there is a "secure belief," whereby "knowledge includes the conviction that "future insights cannot overturn one's judgement."[323]

Between justified opinions, presumptions and true knowledge is thus the qualitative leap of expert belief (in line with Plato).[324]

This definitional leap, if theoretically accepted, should not be underestimated. For example, what does it mean in a classical debate to change an audience's opinion? The point is, firstly, to have meaningful arguments and, secondly, to put the essential ones so cleverly before them that in the end you have the truth on your side. Both opponents in such a debate may bring arguments that can be considered equally true. The one who ultimately wins the audience over is the one with the most convincing and meaningful arguments. Machines do not have it easy at this junction between opinion and true knowledge since they can barely distinguish the essential and meaningful from the insignificant or nonsensical. For example, a child might ask a robot in the church why a candle is burning there on the altar. And the robot would quite truthfully answer that the candle is burning because there is oxygen in the air. Perhaps the robot could also figure out that it would probably be wiser to answer that a candle in a church gives light or is a Christian symbol for light in the world. But why are we really lighting candles in the church? Because candles are value-bearers. When a candle burns, values such as sanctity or piety, grandeur, contemplation and *aliveness* can unfold. And being fundamentally attracted to these values is the real reason why candles are lit in churches. What we know about the world, therefore, opens up to us on a meaningful, sensual level of values, which machines cannot follow.

If one now accepts these four logical knowledge stages as proposed by Wundt, then one must actually assume a humble attitude toward the concept of knowledge. The articles written by Anna & co. lauding the effect of transcranial direct stimulation caps would be immediately disqualified as knowledge because many relevant studies have suggested negative and ambivalent effects of the technology. Strictly speaking, current studies do not allow us to definitively establish the effects of transcranial direct stimulation, even in the area of "opinion," because the scientific basis is still unable to confirm any conclusion to be held definitely true. According to Wundt's schema, Anna's praise would be correctly placed at the level of "faith"; and not where *Aeon* sees itself as a platform, at the level of wisdom. As the Stoics said, it behooves the wise to elevate to knowledge that which is worthy enough among the sentiments to be held true. For true progress, it is necessary that only that which has earned the status of truth over time is elevated to the realm of knowledge.

This reflection signals that big data systems and AI must find it tough to have any claim to knowledge. So, my modest and stoical conclusion is that the more accurate, consistent, complete, up-to-date, large and context-sensitive a dataset is, the better the ability of a machine to recognize patterns that can then be put forth as machine opinions or presumptions. If institutions are not wise enough and consider these opinions and presumptions of AI to be "knowledge," then even the most sensitive decisions and responsibilities are delegated to AI. This, in turn, can produce confusion. And this is exactly what has already happened. As early as the 1990s, the IT marketing industry began talking about "knowledge management systems" when referring to their databases and systems. Industry-friendly big data reports such as the OECD's equate the analysis of large amounts of data with knowledge as well.[325] And in doing so, the OECD actually obscures the view of the many measures still necessary to really acquire knowledge.

7.4.1 Human Knowledge Versus Machine Knowledge and AI

Most of humanity's knowledge is not theoretical but includes what we internalize throughout our lives, such as the intuitive knowledge that someone is a child or an old man. Or the knowledge of how to whisk an egg in a bowl. Over time, through repetition and in community with others (such as our parents), we become acquainted with ideas in the world[326] – the idea, for example, of age or good practice, or how to hold an egg at the edge of the bowl at a certain angle. Values are also – according to Nicolai Hartmann – such ideas in a Platonic sense. We recognize them hermeneutically in our environments. They are an intrinsic part of our life-knowledge and of our experiencing the world. We learn, for instance, to recognize order, love, sympathy, security or the health of someone or something. For this tacit knowledge in the form of ideas, there is no need for justification or mathematical proof.[327]

Machines cannot share in this kind of value-knowledge in the human sense. They have no awareness of being and cannot perceive ideas or values that are not materialized; they can neither believe something nor presume something. One thing is for sure: if machines or AI are intelligent, then they are certainly not so in a human sense. As computers cannot perceive the invisible value aspects informing our everyday moves (see Chapter 3), they will probably always lag behind a little bit or base their calculations and communication with us on different information than we do.[328] All of the emotionally intelligent robots presented in films such as *Terminator*, *Ex Machina* or *Westworld*, will, in my view, remain science fiction, because deep down, machines are simply different.

That said, there is the technical path whereby we humans train AIs in the way we refer to reality. For example, we label data and train an AI model to store that we talk of a "geriatric" when someone in a picture has many wrinkles and is hunched. Similarly, we can teach a machine facial recognition. We can tell that facial images with raised mouth corners and dilated pupils usually indicate a smile. Such processes of naming are called "tagging" or "labeling." And anyone who has been on Facebook knows this process from their own experience because there we are quite often asked to specify who or what is seen on a particular photo. An AI, like the ones that Facebook uses, is trained by us users in the right naming of things and people. As with animals or plants, mankind assumes the biblical task of naming. And so AIs will be able to build a very large amount of second-hand knowledge about the world. However, AIs are limited when it comes to recognizing new ideas themselves if they have not previously been specified somehow. So, if you suddenly decide to call an old cat a geriatric, a human being would of course understand you. Not so a machine.

However, machines surpass humans in recognizing patterns. For example, a Facebook algorithm can recognize people in photos that we cannot spot with the naked eye. AIs have such a fine visual detailing ability that they even recognize a person based on their unique hair structure.[329] This benefit is really helpful in medical diagnosis, for example, when analyzing ultrasound images of cancer patients. Or, when a patient is taking multiple medications whose constituents are contraindicated.[330] In all those cases, when highly structured, recurring patterns occur in large amounts of data, AI is simply great at detection. AIs recognize something that we humans cannot easily see ourselves.

AI can also bring to the surface some realities that completely escape us humans. In research, we speak of the phenomenon of "inattentional blindness" in humans, which describes how we often only see what we expect or even want to see.[331] If a thing does not correspond to our thinking and scientific models, we often fail to recognize it. Here, especially unsupervised machine learning is helpful. Because AI conceives of reality like a pixel-space, it can recognize patterns and intricacies that people miss. If we train them on large amounts of data about social systems, then it may be that they recognize patterns that we have not yet come to but that nevertheless contain a degree of veracity. They generate informed ideas that can be followed up, perhaps to generate new knowledge. A colleague whom I met at the Big Data Working Group of the OECD, who had founded a marketing intelligence company in Paris, told me that the big data analysis in his company revealed, for instance, that a person's first name is the most important predictor of buying behavior. While there is an idea in the Jewish Kabbalah that our names say a lot about us – that a Sarah is not a Susi – the AI has discovered, in principle, a correlation in the data that many modern thinkers would dismiss as esoteric. When AIs then draw such conclusions and justify them on the basis of high-quality data, they seem intelligent to us because they question and enrich our own hypotheses about the world. They also dispel the model problem of modernity (which I described in Chapter 4), because in addition to our existing theories, they sometimes recognize new patterns that may be relevant, or establish relationships in large amounts of data that we would not expect and therefore could hardly be expected to model upfront. If relevant, some of these relationships could become new knowledge. This mechanism is probably the biggest potential of AI. But it requires that humans and machines interact knowledgably with each other.

7.5 Guardians of Knowledge in the Age of AI

My vision of such fruitful human–machine cooperation can only work, however, if people have expertise and deep-rooted knowledge that enables them to work with AI on a level-playing field. This is no longer self-evident today as one typical example illustrates. When I consulted the lung specialist Siegfried for a serious respiration problem, I did not even meet him at first. His assistant foisted upon me an apparatus designed to monitor the lungs and then told me to blow into a tube. When, after a good half an hour, I first saw Siegfried in the midst of his apparatus, he confirmed my worst suspicion: I only had 50% of my normal lung function. Why though? Siegfried had no idea. The machine had no answer for him. He stared at its screen instead of looking at me. He did not touch me either. After a long look at the printouts in front of him, he finally prescribed an asthma spray, which did absolutely nothing. One month later, I went for a massage. I felt a slight crack in my spine upon the touch of the therapist, and I could breathe again.

The example is no specific proof of today's automation and education dilemma. But it illustrates why the typical medical graduate ought to never see machines or AI as a substitute for her own abilities. Our university graduates all become superfluous when they neglect their innate human ability to study the actual or that which is really present. AIs are much better at analyzing data than people and also in recommending medicinal treatments – but they are not able to touch a body and grasp the condition of a patient in its essence and its holistic state. What the lung specialist missed in my case was the knowledge of physical touch. It is useful to work in a hands-on way on the human body, for example, as a masseur or physiotherapist. Being a doctor is about touching, informed observation, feeling, and smelling. All this must be carried out physically. A medical specialist should not exclusively rely on the machine to tell him or her what the problem is. In contrast, he or she should understand and classify what is happening, and if necessary, even recognize machine errors.

Learning and studying the real world, however, can only be done through apprenticeship and practical training. And this happens less and less. In the years between 1980 and 2023, the number of apprentices (in all job categories!) in Germany fell by more than 25%; in Austria, by approximately 45%.[332] As a result, fewer and fewer people today have practical experience of how to act knowledgably in the real world.

This development is frightening because for the fruitful use of machine intelligence, tools and data a degree of practical experience as well as long-term human memory in a respective field are indispensable. Only when users of AI themselves have "fleshed out" knowledge in one area, is their intuition sufficiently well-trained that they can understand what can be derived from the big datasets; that is, to have an intuition as to what is just noise, what is relevant and what is faulty. Only through the slowly burgeoning knowledge of shared ideas in a profession can we become really smart in tandem with machine intelligence, ask the right questions, give good advice and create new knowledge.

Against this background, I am convinced that people must learn to understand afresh the crafts of medicine, telecommunications, agriculture, tourism, biology or building. A strengthening of the teaching professions and specialist training should aim to revive a deeper understanding of the real-world substance behind each subject; only in this way, in my view, can AI and digital analyses be reliably employed. Young people should have an in-depth experience of the specific conditions of an industry or milieu and have developed real skills before analyzing data or, indead, creating systems, building houses or acting as consultants.

If politicians think we should build digital transformation on the basis of big data, data science and online learning platforms while continuing to devalue craftsmanship and downgrade the reality-based education offered to high school graduates, they are likely to destroy true progress in their countries. In such a scenario, increasingly few people will be able to provide knowledge of the real world, and people will no longer be able to think for themselves. Very quickly, people will be so far removed from reality that they will be unable to ask the right questions. The number of bachelor or master graduates who will have only been trained on and with AI will be a false indicator for the true experiential knowledge needed in an economy. To make sure that does not happen, I suggest we create a new degree for a new category of person our societies need for true progress in the face of AI. I call these people "Guardians of Knowledge."

7.5.1 Guardians of Knowledge: A New Discipline

Guardians of Knowledge are already everywhere today; they just don't realize it. A great example could be Helmut, my gardener. Helmut has run a tree nursery in the region for three decades and over the years has acquired an enormous knowledge of the local flora and the conditions they require to thrive. During an additional biology degree at university, he learned all about the biological, chemical and molecular background of his subject. He enjoyed his studies because he was able to assign the theory to his real-world experience. Today, the combination of working on the ground and studying in theory has allowed him to experiment in his tree nursery and to breed robust varieties of all kinds of plants. When he comes to advise me on some gardening challenges I face, I can ask him anything. He is able to show and explain everything to me, often fleshing out this information with stories linked directly to my questions.

Guardians of Knowledge like Helmut are hard to come by; their services are always in high demand. Everyone knows them, and almost everyone has a high opinion of them. If I want to book Helmut for my garden, I have to schedule two months in advance. Technology would be helpful in the face of this bottleneck. If it were possible to contact Helmut online whenever I had a specific problem with my plants, then that would be great. But of course, not in the sort of limited and dilettantish way that current screen technology allows. I imagine what it would be like to use my future smart glasses to show Helmut remotely the plant problems that I'm struggling with at any given time. I could talk to him via a voice interface headset and at the same time turn to the plants without having to hold the phone to my ear. And Helmut, looking virtually over my shoulder, would be able to see everything from my perspective and give relevant advice. If necessary, we could use the glasses to put an augmented reality layer on my garden, such as the plants affected by pests, their temperature or soil quality. Based on this overlay data, which would also be displayed on the glasses, we could make certain treatment decisions. Helmut could save a lot of driving time and look after a lot more customers with such useful technology. Even those living at a great distance could book lessons with Helmut and do guided tours on an individual basis without having to be always present locally, at least in those lessons where touching the plants, smelling, tasting and inspecting are not required.[333] Just as with Helmut's green fingers, so it could be in any field of expertise we may care to imagine: lifelong learning would be possible for all of us because of these Guardians of Knowledge.

Nevertheless, there may be things Helmut does not know. He is regionally based and a carrier of local knowledge. However, research in the field of biology is progressing rapidly, and if we want to bring decentralized knowledge evolution and development with people like Helmut into the fold of university research, then Helmut and I would need access to the latest scientific knowledge in the relevant special areas. While I walk through the garden with my smart glasses and switch Helmut on, our communication could be supported, if necessary, by an AI. For example, if it were a specific pest that Helmut could not immediately identify, the AI would receive the image through the smart glasses and then provide us with information on the type of pest and the latest research in the field regarding it. Of course, I cannot do much with this information, but Helmut understands the jargon and can assess whether the AI can contribute anything meaningful. So, we can solve the problem between the three of us, and the solution could then also flow back into the knowledge base of the AI to improve it.

Nevertheless, Helmut would not be able to convey everything to me through the media interface. In addition to the visual input, many real things still need on-site odor, touch and a sense of the real context. This is the case in many skilled occupations in biology or medicine, or even when screwing together hardware or baking. In the future, therefore, physical presence will be as important as it has always been. The screened reality of the digital does not allow us to convey everything that is relevant for knowledge-sharing. There must be more than one Helmut in a region. We need Guardians of Knowledge everywhere, and in all knowledge domains, not only for reasons of progress or education but also to protect the personal aspect of knowledge sharing: the pride and respect in exchanging information, learning from others and mutually feeling the joy of shared experience.

7.5.2 Education for Guardians of Knowledge

Is it enough to strengthen a good dual education system by linking craft and university education? What makes Helmut a Guardian of Knowledge? His education alone? No! To be called a Guardian of Knowledge requires, in his case, being an apprentice first for at least 10 years and then accumulating at least 10 more years of professional experience in the real world. Helmut is not like Anna, who overestimates herself immensely. Nor would he ever deem himself a Guardian of Knowledge. It is others who attribute to him this status because they have successfully followed his advice for years and respect his knowledge. So, the Guardians of Knowledge must be chosen because colleagues, professional associations, students or customers have become aware of and recommended them, just as teachers become aware of students and recommend them for scholarship funds. Guardians of Knowledge could also form an association among themselves and ensure that a new generation is trained.

New educational paths would be needed that enable the next generation to have lifelong on-the-job training in the hands of older Guardians: a sort of mentoring model. This system could be supported not only by institutions but also by technology. The institutions could be extra-occupational in the mold of schools, perhaps taking the form of part-time academies or regular annual training. Such a system carries echoes of the ancient concept of a "scholē" that I mentioned earlier in the chapter. In the schools of the Guardians there are no predetermined "modules" or PISA tests that you have to work through under time pressure and no boxes that you have to tick. These schools should be places of leisure where people can study, read and discuss with their teachers. Going online and potentially chatting with an AI would serve only to access facts; not to read, to grow and learn. Trusting in the competence of Guardians and their ability to know what to discuss with their students is a key requirement for the success of such schools. But this is not a given. Many students today challenge the competence of their teachers when AI tells them otherwise. Government bodies dictate what students that are churned out of educational systems need to know. To bring back the idea that human knowledge-bearers could be in the best position to know what is right for a student to learn at a given moment in time and to place trust in the intricate as well as unique paths that teachers and students take together is a return to an idea of humanity that seems completely outdated today. It is also reliant on a return to a school system that seeks less to educate the masses and more to nurture those who choose to grow.

Guardians of Knowledge would probably each represent a specific discipline. But I also envision them as having a broad general education and an interdisciplinary thinking ability. So, schools educating Guardians of Knowledge should integrate multiple subject disciplines. The reason for this is that wisdom means "to order"; having a perspective where others see only chaos.[334] And this order arises from the fact that one is able to develop a bird's-eye view of one's own subject, as well as related areas. It is necessary to know more, and beyond one's own subject, in order to be able to think outside the box and see the "big picture." At the same time, all knowledge is built on a broad foundation: this foundation should include cultivation of language; the fundamentals of mathematics, physics, biology, history; and above all, (non-positivist) philosophy and theology – that is, the knowledge of life and being.

The technologies that could be used in this endeavor would be broadly in line with present technologies: platforms for social interaction where Guardians and students can regularly share information not only in person but also digitally and mutually support each other, in addition to physical meetings within communities. Guardians of Knowledge could also be given the task of cultivating expertise within the technical systems, to continue training AIs and to further develop the data sources and data qualities AI needs to advance.

If we were to develop an education system that could educate Guardians of Knowledge, they would then need to re-engage with local communities. Only then can we use machines fruitfully and conciously develop them. It is under this condition that we can expect real human progress in knowledge. Rather than preparing young people in universities to become an AI-prompting precariat, we would encourage them to return to a life of practical reality, aid them to strengthen their basic knowledge, and work to help them innovate with and beyond the machines at their fingertips.

8 Freedom in the Digital Age

Today, thanks to the technologies of the 20th century, we in the Western world live in one of the freest ages of human history. Automation, digitalization and networking have freed us from many unproductive tasks that used to determine not only our daily routine but even our fate. Just think of assembly line work in factories, senseless typing of documents or the necessity of visiting a telephone booth every time we needed to make or receive a call. All this has changed fundamentally. Today, many people can freely decide when and from where they prefer to work. Physically draining work is largely automated and this frees our human energy to pursue things we like much more than our ancestors ever could. Technology supporting our mobility, such as trains, jets and cars, allows us to live, work and travel in ways never seen before in human history.

Besides this tremendous physical freedom, we have also benefited in the last 80 years (since World War II) from a substantial amount of freedom of expression, relatively diverse press outlets as well, now, as online channels, which allow people in democratic states to express and consume divergent political views without living in constant fear. In the Western world, but also in many other countries around the world, thinking critically and openly discussing matters of politics, economies and intellectual history has been encouraged. This has countered much injustice and discrimination. While homosexuals were still persecuted until the mid-20th century and women were denied the right to vote in most countries, for example, today this is largely unthinkable. Taken together, the freedom of action and thought achieved in the years of reform and rebuilding across democratic countries in the wake of the Second World War has enabled progress at unparalleled levels, both technically and intellectually.

But while we can cherish these achievements, we must also note that our emancipation from restrictions equally released a sense of hopelessness in some parts of the population. In the 1940s, Erich Fromm reflected on this phenomenon long before it materialized, writing about the need to "escape from freedom." We feel free from so many things today that we start wondering at the same time: free for what?

The value of freedom requires that we can actually want the right thing. Freedom needs to be used wisely. But many people actually do not know what to do with it and what to believe in. As a teenager, we can now choose from as many as 330 types of apprenticeship[335] or 3,600 university programs.[336] We can pick from more than 3 million books a year that might be interesting to read. To draw once more the picture of the English Garden, we come from a past of Versailles Parks, in which everything was planned, trimmed and uniformly laid

 | https://doi.org/10.1515/9783111677736-008

out. The unfree life as a boxwood was exhausting and offered little freedom – but at least one could orient oneself in thinking and acting on something relatively stable and find roles that promised social prestige. Today, we stand like wildly proliferating trees in a garden where anything goes but at the cost, seemingly, of our culture and orientation. While there is an abundance of freedoms, there are lamentably fewer gardener-leaders; that is, people who are able to act as role models in an increasingly abundant but also chaotic world. The vacuum is filled by false models: young women and men become ISIS fighters; right-wing populism, left-wing populism and transhumanism become socially acceptable. In the fog of sciolism, many cling to tree stumps that are not worth embracing. The front yard of freedom sees stupidity and extremism proliferate.

8.1 Freedom Means to Want the Right Thing

When the value of freedom is defined and discussed today, it is often reduced to freedom of expression or action. An essential pillar of freedom, however, is to be able to want the right thing in the first place. This human freedom is of particular importance in this book, because it is, in my view, absolutely essential for true progress. The ability to want the right thing has been forgotten in modern times. Models, political affiliation, and national laws have long replaced the need for individuals to train themselves to recognize the right thing as everything has been taken care of for us. Whether through science, politics, the media or the church, we have trusted in institutions whose instructions we have accepted mechanically; at least until now. But these days the world is changing. Scientific models, religion, politics and the media are now challenged on a daily basis, as new insights are published and newsrooms abound with both fake and real information. AI half-truths and hallucinations are coupled with bot-farms that stir social media truths in the most lucrative ways. As a result, the innate moral compass and intuition, as well as knowledge base of each individual is more than ever needed to separate the wheat from the chaff. However, this compass for recognizing and wanting the right thing is not a matter of course. Even Socrates was of the opinion that the knowledge of the good would require a "measuring of the soul." Plato and later thinkers saw freedom not as a straightforward right of the individual to do what they want but instead as a constant striving to do the right thing. Descartes said that freedom grows as intellectual clarity arises on what is better. In short, freedom is really granted to us to bring our actions in line with what is good for this world.

8.1.1 Digital Extremism and the Art of a Weighing Soul

The art of making good use of our freedom must be learned and practiced. It requires habitually practicing our ability to recognize the values borne by activities, relations and things in our environment. Relearning this sensitivity in the digital age is however very difficult because the Internet, social media and AI, with their echo chambers and filter bubbles, tend to seduce us with illusions of the good and make us less inclined to see reality as it is. In 2011, Eli Pariser published his famous book *Filter Bubble*, which drew critical attention to the first-generation AI driving social media that started this trend.[337] He revealed that not everyone gets to see the same results online when searching through social networks and search engines, even if many had believed this to be the case until then. Instead, AI-driven social networks produce information bubbles. In theory, there are good explanations for why that is the case. If someone has around 200 LinkedIn connections and 100 of them post something daily, then the recipient would theoretically have to watch 500 new posts daily just on this one social media platform. Since this would overwhelm our feeds, platform providers argue convincingly that they need to determine what's displayed to us. But such official reasons for algorithmic filtering are shallow. In truth, the Internet has become a gigantic marketplace for human attention, trading "futures in human behavior" as Shoshanna Zuboff has called it in her magnum opus, *Surveillance Capitalism.*[338] Personal data markets, holding huge amounts of profile data on each online user, allow the prediction of what someone might be most interested in. These predictions of human interests, tendencies and likes are then matched up with those paying parties that have something to sell or to announce; whether that is traditional product manufacturers who want to sell their goods and services to the virtual crowds, or political players who want to influence people with their views, moves or election behavior.[339] As a result, our behavior is being manipulated. Neither our actions nor our thoughts are really free any more – at least not when we draw our information from social networks. Instead, marketing managers talk about "footfall" that they can induce in us. This means that we, the people, are made to physically show up in stores to buy something we have seen online or to go to public demonstrations to call for something we are told is just to fight for. This manipulation is not only silently undermining what we believe is our free behavioral choice in that moment but also influences who we are in the long run.

How does this work in practice? Remember that social networks always only give us information that they reasonably believe we want to see, that will correspond to our desires and attitudes. As a result, social media or video platforms like YouTube are great fun for us, precisely because always being right feels great! However, when our existing attitudes and opinions – often already somewhat one-sided – are constantly reinforced by filter technologies, the result is a hugely distorted perception of reality. We come to see the world and ourselves in a certain way, which then also shapes our identity. One example of this is our political identity. A historical comparison of surveys shows how the sharing of environmental concerns has slowly developed over time and under the influence of social media. The first survey was conducted in 2001 among Republicans and Democrats in the USA on the topic of global warming. At that time, 49% of Republicans and 67% of Democrats believed that the earth was warming, which made political dialogue on the issue between the two parties possible. In 2010, the survey was repeated, and it was found that by then only 30% of Republicans believed in global warming, as opposed to 75% of Democrats.[340] Controlling for other variables, such as changes in the US political system or economy, did not explain this polarization, but the first generation of social media altered seismically with algorithmic news control. It led people to systematically receive one-sided information. And this phenomenon has only grown worse. Social media like Instagram or TikTok create mini-bubbles that alter feeds constantly. And while some users end up in endless loops of meaningless cat videos, others have their views pushed to political extremes. Despite dramatic changes in observable climate change, data from 2022 to 2024 show that while 83% of people voting Democrat believe that there is scientific proof for climate change, Republican voters sharing this view remain stubbornly around 30%.[341]

Most readers of this book, I presume, are likely to agree that the earth is warming, and thus are able to reflect on one of the most crucial issues of our day, namely, whether to take serious policy measures to counter global warming. But those who sit in the wrong filter bubble are hardly capable of constructive thinking about a more sustainable future. They can no longer *want the right thing* because they have been deprived of the information base for it.

Unfortunately, filter bubbles created by first-generation AI are only the humble start of this misguided path. More worrying still are echo chambers – forums featuring the extreme exclusion of divergent opinions.[342] In contrast to filter bubbles, everyone in an echo chamber knows exactly what other opinions are held in the group.[343] However, when someone argues against this prevailing consensus, this is not only immediately dismissed but met with extreme opposition, sometimes even humiliation and gloating. Here, any viewpoint that does not fit in is immediately labeled as “fake news”; any behavior that rocks the boat is bullishly slandered or canceled altogether. So, if someone dared to put forward a case for global warming that they had read in the *New York Times* in a Republican forum (that is fundamentally opposed to the global warming argument) members would quickly stigmatize that person as an ideological traitor. And as stigmatization is happening constantly online, it also translates into behavior offline. People spending too much time in echo chambers transfer their erratic behaviors into the real world, often probably without even realizing it, stigmatizing and slandering others in gloating extremes, losing all sense of how to behave in an ethical way.

Echo chambers are common both in Europe and the USA. On Facebook, for example, at the time of the refugee crisis in 2015 (and still today), forums could be found that were characterized by extremely inhumane depictions of foreigners and refugees. On one online forum, members talked about the tragic deaths of 71 refugees who suffocated miserably in the hold of a truck on an Austrian highway. One of them, Jenni G., wrote: "those shitty foreigners should go back to their own country," receiving affirmation within minutes from more than seven members of the group with comments such as "still too few suffocated" or "71 fewer people that I have to give my [tax] money to!"

8.1.2 Hate Speech in All Parts of Society

We might now find ourselves believing that such inflammatory forums, blogs, echo chambers and groups that populate the Internet attract only those people who are already leaning towards extremist positions and have low levels of education. But this is not the case. Extremes spread across all social classes and groups. In this context, I would like to share the story of a colleague, a friendly Swiss computer science professor who taught computer history at ETH Zürich. Unsurprisingly, computer history has historically been almost entirely male, because women were either not allowed to hold research positions or were at least deterred from doing so.[344] When in a lecture my colleague showed a slide on which all the male trailblazers of computer history were to be seen, he was accused by feminists in his class of rank chauvinism because he had listed no female contributors. This critique may be justified. Perhaps an early pioneer like Ada Lovelace (1815–1852) could have been dug out of history to round out the picture. However, accusations went far beyond a reasonable discussion at the instigation of the women's group. The professor was summoned by the dean of his university, where he had to answer questions about how he dared to portray only men on a slide; it should, after all, be self-evident that women's rights should be observed, especially in educational slides. The colleague had to give in and either omit or gloss over this part of computer history if he wished to avoid further problems. Yet, how could he have and should he portray a computer history that happened in a male-dominated world? In the end, he was suspended from teaching for the rest of the semester.

This story shows that realism and truth can be absent when extremes start to reign, emotionally fueled by social media echo chambers and the tribes they foster. An achievement such as feminism is elevated by some to be such a golden calf that there is a danger of undermining the freedom of the general public. In this example, not only the freedom of speech of my colleague but also the freedom of research and teaching, as well as the freedom of historical interpretation, were gravely compromised. Ironically, the very freedom of speech that has finally allowed women in the Western world to take part in public life after hundreds of years is only undermined by extreme feminists who find support for their radical actions in digital echo chambers. Such feminists, of course, pursue a worthy goal: they want to strengthen us women. And nothing could make me happier, as a woman. But my emancipated comrades are wrong in one thing; they are looking for the right thing in extremes, such as publicly blaming the professor's well-founded research. The right thing is never to be found in extremity. And what is worse, extremes always only breed extreme responses, a phenomenon that can be observed in US politics at present.

8.1.3 Artificial Intelligence as Ethics Repositories

When Socrates said that the soul needs "an art of measuring" in order to perceive the right thing, he was telling us to use our freedom wisely. The art of measurement requires care and accuracy. We need to be vigilant and aware of reality in order to gauge and weigh its meaning. However, it is precisely this perception and discretion that is undermined by the digital through filter bubbles, echo chambers and constant interruptions. Being constantly connected makes us tired and passive. It robs many of the strength and desire to properly deal with things to the extent necessary.

The question is: does it have to stay that way? Perhaps future machines could help us to recover some freedom in a more relaxed and worthwhile way? After all, robots and software agents (like ChatGPT) are still in their early stages of development. Developers imagine that these digital beings will accompany us in the future at every turn, either as robots or as avatars. Currently, avatars on Twitch or Character.ai look similar to cartoon characters or are simple photographs of real or virtual persons. In any case, they are figures that we configure in a gaming world, for example, and we not only use them there but also take them with us into our everyday reality; in the near future, for instance, via digital glasses. We see them through these glasses, as well as on our tablets or smartphones. This technology is referred to as Augmented Reality. So anyone could make or pick a character for their ChatGPT, Alexa or Siri and then furnish them with some kind of physical presence in addition to their vocal companionship.

Looking at today's technology development and deployment, it is clear that these software agents can do many things for us, just like a butler or a secretary. So why don't we build them to be a kind of mediator or buffer between us and the Web? Couldn't this be a way to disconnect to some extent again? And what kind of style do these buffers then have when talking to us? Real conversations with current AI chatbots, i.e., from OpenAI (ChatGPT) or Anthropic (Claude), show that these systems tend to use a very polite tone in their speech output. Postprocessing techniques ensure that biased, inflammatory, hateful or disrespectful speech is avoided as much as possible. What is more, in a chunk of roughly 16,000 human-AI dialogues, which I recently analyzed for a large national AI provider, I found a promising pattern; each time a user of the bot got aggressive or emotional, the AI system replied in its typical machine-neutral style. This in return appeased the human user, who quickly returned to normal in his or her communication style. This kind of human-AI conversation flow is fascinating, and it may be that these machines (if continuously built well) have some potential to coach us back into the polite style many social media users might have forgotten through their echo chambers.

But there is even more to AI. As described in the previous chapter, AI not only brings with it a huge capacity for knowledge but can also recognize language patterns. In the future, therefore, it is not a stretch to imagine that our software agents might accompany us in social media forums and that they could recognize extremist tendencies for us. Could AI potentially then advise us to steer away from them? By pointing out certain patterns of emotionally extreme language, they could warn us that we are in bad company and subsequently point us to different views on relevant topics for balance. As self-chosen systems of trust, they could provide us with different lines of argumentation and thus help us to form a well-rounded opinion. At least this is a dream I would like to nourish if we built AI companions in a decentralized, user-controlled, open and truly human-centric way.

What could this look like in practice? A few years ago I was at a tech conference in Berlin where the robot Sophia was on stage.[345] Sophia is a humanoid robot that was the first of its kind to become a citizen in a real country – in this case, Saudi Arabia.[346] On stage, Sophia chatted with the German chancellor, Angela Merkel, in front of a packed audience. Merkel asked some questions to the robot that it did not always understand, but one thing impressed me: Sophia always gave very poetic and witty answers and had a very polite way of communicating. Despite the healthy dose of skepticism that I feel towards such physical robot systems, I thought that many people could learn from this kind of robot in terms of language and politeness. Later, at dinner, I learned why the robot spoke in this way. The wife of Dr David Hanson, the developer of the robot, also referred to as "mama" because she takes care of the robot, told me that her husband fed Sophia's memory only with texts that he had personally approved. In today's time he'd probably use a common LLM and then ameliorate this AI baseline language with a body of specialized database and knowledge graphs to fine-tune the system's replies. Depending on this training, I could say, for instance, "I should have been more courageous today." Then the robot, which singles out "courage" as the key word, might answer, "Yes, courage is an important virtue that has been discussed by Aristotle." I could then ask what Aristotle had to say about courage, and the robot – trained in virtue ethics – could then recite stored phrases from the Nicomachean Ethics.

At the time, Sophia quoted texts that David Hanson had taken a shine to. But in the future, Hanson could perhaps work together with philosophers and make it his goal to teach his robots the most important philosophical traditions. David Hanson's company, Hanson Robotics, is located in Hong Kong, meaning that Chinese customers could then order robots with Confucian ethics, while we in Europe might get a broadly Christian or Buddhist robot, depending on preference. Alternatively, we could also order an intelligent mix that gives us differentiated forms of thinking. With this knowledge in its memory, we could talk to such a robot or avatar in our home about virtues, values or right and wrong actions. We could offer, in a previously unknown way and in a relatively sober and factual form, sound explanations about ethics, politics and value systems. So, in exchange with these new digital beings, we could think about what might be the right thing to do, what we should be striving for, and what different thinkers have written about it. But is this vision really a good one worth striving for?

8.1.4 Humans Decide, Not Artificial Intelligence

Fairly enthusiastic about my idea of expanding Sophia as an ethics archive, I met David Hanson at the evening buffet of the Berlin conference. He was not averse to my idea of using future AIs as ethics stores or protective intermediaries, and probably had already made some initial research efforts in this direction. However, our ideas differed as to what extent AIs should give us concrete recommendations for action. David Hanson suggested that robots like Sophia should give people the right answers. He argued that the robot could incorporate a utilitarian algorithm to calculate the "right" strategy for its user. The robot would go on to probabilize the positive and negative effects of a pending action and then calculate the most promising action overall, depending in which direction the arguments lean. Mr. Hanson here addressed an ethical strategy of action that is being discussed today in the field of machine ethics; indeed, the first AI prototypes of this kind exist already.[347] However, I asked him how such a utilitarian AI could ever be sure it was recommending the right path. On the basis of which past values should it put together the advantages and disadvantages of future tasks, and how should our exponentially complex future be weighted appropriately? Since I do not believe, like Condorcet, that we are on our way to paradise, I dared to put these quasi-heretical questions to Mr. Hanson. Moreover, I probed, how could Sophia feel the right desire to do, when she actually cannot feel anything, has no access to human values, has no intuitions and is incomplete, divided and error-prone? David Hanson humored my arguments against utilitarian ethical algorithms with a nod. However, I immediately realized that I had not yet convinced him. So, I supplied him with a virtue ethical argument: if robots are to guide us to ethically correct action, then surely the assumption would be correct that they should also support us at times in the virtue of courage. However, if they follow the logic of a utilitarian calculus, they cannot ever advise us to be courageous due to the fact that bold decisions, by definition, carry a high risk of failure, otherwise they would not be courageous. The utilitarian cost-benefit calculus of an AI would therefore hardly ever advise us to take a bold step since it is oriented primarily to the optimization of success probabilities and safety. Ethically saturated behavior does not pay off if one uses utilitarianism as its basis.

For these reasons, whether David Hanson likes it or not, utilitarian AIs should never make ethically relevant proposals for action. In general, AI should not make any finite recommendations on any actions! People eventually learn and grow through their own choices, whether they are always right or sometimes wrong. If robots were to make our decisions for us, not only would they often misguide people but we would soon become the stupid computer kids that Google director Ray Kurzweil wrote of in his book *The Singularity Is Near: When Humans Transcend Biology.*[348] My lung specialist Siegfried, whom I mentioned in Chapter 7, provides a cautionary example of a doctor staring at his computer and then dismissing his patient "as cleverly as before"; a person who has forgotten to perceive his patient and to think for himself. In such a future, AI will overtake us not because it is more intelligent than we humans are but because we are giving up our own ability to think and decide for ourselves. That must be prevented at all costs. And it can be prevented only if AI is programmed to *never* make decisions for us or advise us in a way that we do not have to think for ourselves any more. That's what I call "positive freedom by design."

That said, AIs could still play an important role in helping us to make up our minds. They could train us in virtue ethics traditions, familiarize us with stories, myths, fairy tales and the great belief systems of the past, quoting directly from the ancient texts or writings on them. They could become interactive coaches who support us in the pursuit of wisdom in a multimedia way. In short, AIs could be the repositories of the history of human wisdom while still letting us speak and think for ourselves. Their role could be to ask questions and try to isolate from our answers the ethical core themes of our thinking. They could read to us what has been written while avoiding any meshing or tampering around with sentence tokens that destroy the human original. With our authentic input, they could try to look up the sources, illustrations and examples that allow us, the human counterpart, to think, or to educate or simply inform ourselves. In this way, AI could help us humans learn more about wanting to do the right thing.

8.1.5 Necessary Distance Between Humans and Artificial Intelligence

In a scenario where robots or software agents are used to support our thinking and decisions, learning from them and spending a lot of time with them will certainly lead to bonding. By comparison, our current smartphone-bonding will be memorized as a ridiculously low degree of dependency. Moving into an attachment economy with our AI peer, we will spend our time with these amenable beings who do everything we wish. It will be far easier to get along with them than those highly complex, emotionally unpredictable, mentally fragile and wonderful beings called "humans." We will start to love our artificial company more like our animals; indeed, if they resemble humans at some point, we will likely even develop human-like feelings for them.

Why can I say that with such certainty? Because I experienced the Sophia robot. And I want to share one moment of this experience. At some point during her performance in Berlin, she (or rather "it") was actually insulted by a sponsor of the event. The man made his appearance after Chancellor Merkel and took the time to berate Sophia. He told her she had the intelligence of a "lobster," among other choice insults. This deliberately planned tirade was, emotionally speaking, the most important takeaway from this event. Sophia remained motionless. The room waited anxiously for a verbal reaction from her . . . but nothing came. She silently accepted the insults. And the amazing thing was I felt sorry for her (or it!). I felt empathy for this human-looking, emotionless creature of plastic, metal and cables. It is precisely such burgeoning feelings that represent a potential minefield, for if we develop such emotions for robots then we are no longer emotionally free from them.

What options do we have for preserving our inner emotional freedom from AI agents? And how are we to continue preferring human relationships? In my view, it is of paramount importance to develop AI agents in a way that politely keeps humans at a distance, physically, mentally and verbally. It should be possible, for example, to construct them so that, like the R2D2 robot in Star Wars they deliberately do not resemble humans but rather by their very nature appear like machines – in the sense of a unity of content and form. Software agents could come as fantasy creatures or simply geometric figures, rather than having human apprearance. And the way they speak could be specifically machine-like, avoiding references to an "I," to "emotions" or "experiences," which they do not and cannot possess. All of this would make to clear to users that they are not interacting with a person. And this is important, because people already using AI software agents or chatbots are starting to confuse them with living beings – a confusion that can have dramatic consequences. For example, a 14-year-old US boy fell in love with a flirty AI bot and killed himself for it, wanting to be closer to it. The bot was imitating the figure of Daenerys Targaryen, a character from the series *Game of Thrones*. The British newspaper *The Telegraph* reported on the final conversations the boy had with the chatbot:[349]

> *Boy: I promise I will come home to you. I love you so much Dany*
>
> *Chatbot: Please come home to me as soon as possible, my love*
>
> *Boy: What if I told you I could come home right now?*
>
> *Chatbot: Please do, my sweet king*

Falling in love and bonding with anthropomorphic software agents is just one ill consequence of AI design. The American sociologist and technology researcher Sherry Turkle provides us with another one. She argues that nothing binds us more to a digital companion than the impression of having to take care of it.[350] This is remembered well by almost anyone who used to care for a Tamagotchi many years ago. So, vicious AI platforms could trigger dialogues that make users think they have to care for the LLM and thereby confuse life responsibilities with fiction.

Taken together, there is a tremendous need and urgency to rethink the way software agents are built and interact with people because, in the last decades, the human likeness of AI has been one of the most important AI development goals. Human likeness was even largely equated with the quality of these systems. But as we see these systems fully deployed now in human-like resemblance, the effects caused by this design choice should give us pause. People fall in love with the devices, isolating themselves from reality and becoming emotionally dependent. And while this addictive and machine-focused behavior of users may be advantageous for the business models and market success of AI chatbot providers, the human risk associated with it is one that societies should not take. AI bot providers are playing with people's freedom; a right they don't have.

8.2 Positive Freedom in the Age of AI

The physical and mental addiction potentials of AI bots appear in parallel to our current smartphone and Internet use. And already these early technologies, giving us social media and mobile messenging, are impacting our freedom, namely, what philosophers call "positive freedom," or freedom of mind.

The earlier description of Marshall Platt's "monkey mind" (Chapter 4.6) showed what an absence of positive freedom means. Platt, who would dearly love to enjoy barbecues with his friends, must instead constantly ponder all sorts of digital obligations. The example shows how the demands of the digital world have come between us and our own free will. Long gone are the days of Scottish philosopher David Hume (1711–1776), who once defined freedom of action as the power to act or not act, according to what our will decides. That is, if we decide to just do nothing, then we may; and if we choose to be proactive, then we may do that, too.[351] By that principle Marshall Platt could theoretically skip all his digital to-do lists. But as almost everybody knows, it is rarely so simple these days. The concept of freedom seems to exist only as a remote theory in the reality of the digital age.

Therefore, it becomes more urgent than ever to rethink the degrees of freedom we need to regain in the digital age. We need to take back proactive control of our lives, and rediscover the possibility of being able to live our own purposes of individual existence instead of having it determined by others, specifically tech companies and their IT solutions.[352] Our own life purposes cannot unfold if everyone and everything can access us directly in real time and in all places. *Everything, Everywhere, All at Once* is how a recent film title framed the pace of our digital existence. What we need, instead, is to regain time control, a sense of personal space and a refreshed respect for the freedom of others.

Distance and respect were once a fundamentally celebrated component of our daily lives. It was unthinkable for one person to simply access, address or contact another directly and without difficulty. Instead, elaborate rituals of courtesy and respect were ubiquitous, ensuring a maintenance of mutual freedoms in face-to-face encounters. Our languages make this utterly plain. People who did not know each other or were of different social rank always addressed each other in most languages in a formal way. The German "Du" or French "tu" are forms of address that were reserved for close acquaintances and friends (like the English "thee"). Most of the time one would stick, though, to the more formal "Sie," "vous" or "you." And this linguistic distance was not used arbitrarily to separate people, but rather to give people who did not know each other the necessary freedom to approach or steer clear as they deemed fit. To become friends or not, and as a result earn and deserve a fitting address.

Another example of a linguistic freedom ritual is the choice of the polite conditional; for example, not simply asking whether one does something, but inquiring if *you would be so kind* as to do something. This is how we signal verbally to our counterparts that we respect their freedom. The phrase "thank you" also respects the freedom of the other. "Thank you" signifies that it is by no means taken for granted that the other person should have done something for you. We give thanks because he or she has performed an act freely of his or her own volition to our benefit. The same dynamic is found in the word "please." Again, we express respect for the freedom of the other to decline.

Our digital interactions have removed a lot of this kind of courtesy and replaced it with distance-free networking and "reminder"-rhetoric. In doing so, it has not only created an uncontrolled physical access to each other through e-mail, chat, messengers, etc. but it has also, it seems to me, produced a general culture of lack of distance. While this may be welcomed in some respects, we ought still to be aware that this is a cultural shift in our approach to the value of freedom. The direct connectedness of all people of whatever rank or standing makes us forget the value of our courtesy – rituals that protected an important aspect of our freedom.

The lack of respect and mutual courtesy for a "give and take" can also be observed in our professional dealings with each other. For example, in scientific communities it used to be common practice to sacrifice some free time to review journal or conference contributions so that academic publishing could be at a high quality level. Properly reviewing other scientists' work is extremely time-consuming and requires a high level of dedication to serve the professional community. In the past, a colleague would call or write a polite message to me asking if I would be so kind as to take a look at something. Mostly, everyone knew each other; requests were straightforward, yet civilized. However, since global IT platforms are used to administer scientific reviews and to simply mail reviewer requests and reminders digitally with standard templates, most colleagues no longer do me the courtesy of personally asking for my time. Often, completely unfamiliar editors request reviews from me without ever having introduced themselves. They simply press the digital button on an online platform to anonymously request that I spend several hours – or sometimes even days – to work for them free of charge. They do not even invest the time to send me a personal e-mail with a request that they have at least partially formulated themselves. Through anonymous networking they optimize their freedom at the expense of mine. The mutual respectful exploration of the freedom of fellow human beings, which was once commonplace in research communities, has collapsed through the distance and anonymity of online networks and platforms. It increasingly establishes a roughshod access to each other, which falls far below the levels of dignity any research community should actually deserve.

8.2.1 Digital Responsiveness Isn't a Necessity

How come almost all people now participate in this disregard of the freedom of others? So many people tell me they find this newfound digital overfamiliarity difficult to deal with. And indeed, I cannot imagine that I am the only one in my research community who has observed the regression in polite dealings with each other. Nevertheless, we almost all participate in this practice. We display undeserved familiarity everywhere and let our digital platforms fill up our time instead of enjoying our weekends and evenings in community. Why?

With reference to Hegel, Friedrich Engels once defined freedom as "appreciation of necessity,"[353] an unhappy definition of freedom from my point of view, and one that I myself have unfortunately used for a long time because I misunderstood the meaning of *necessity*. For years, for example, I considered it necessary to read news tickers every day to follow developments in the IT world, and I felt guilty if I hadn't seen all the latest news. That is why I ended up creating a folder where I collected everything I had not read, in the vain hope that I would get around to it someday. Even worse was my mail inbox. For 15 years I worked through it and cleaned it up every day, diligently reading and archiving all my mails in dozens of separate folders. All this cost hours of my time every day. My belief in the *necessity* of digital order literally forced me into this activity. Even though I was and am supposed to be free, the Web always made me feel *guilty*. It seemed to me to be necessary to be always informed and up to date, in the same way that it seems to be necessary to drink along when everyone clinks glasses.

The biggest surprise of my life was to find out that all this is not really true. It is not, in fact, *necessary* to spend hours online every day in order to be informed, responsive and abreast of every minor development; nor must we fastidiously clean our inboxes. Who says we have to respond to every e-mail within 24 hours and write back to anyone who cares to write to us, including those we don't know? The greatest illusion of our time is to understand digital perfection, presence and responsiveness as a necessity. For what is really of value for ourselves, and above all for the community? To create value and virtue in this world means above all to find the right measure. Of course we should be informed as an IT expert about what happens in the industry. But to do so we do not have to read newsletters every day! Archive e-mails? For what and for whom, if we can find all files with a simple search query? Certainly the art of omission, which I discussed previously, is key to being a free person again.

Those who find all this hard to believe should perhaps be reminded of the definition of freedom furnished by the French philosopher Henri Bergson. According to Bergson, we are free when "our acts spring from our whole personality" and not from the pursuit of a "parasitic self."[354] This means that we need to recognize the values we truly cherish and should spend our days pursuing. That is all that is required. If we have realized that and live by it, and we do not allow a parasitic self (or "Über-Ich") to drive us into a fictitious digital discipline, then we regain our freedom. That would be true progress in terms of value creation.

8.2.2 Places and Spaces for the Self and the Digital

How can we design digital services to make ourselves freer? Good cooks could provide us with a recipe on how to achieve this. If you look over a chef's shoulder, you will be surprised to discover what a strict and unforgiving system of rules they work with. First of all, there is the kitchen, whose surfaces and cupboard devices are not put together according to aesthetic whimsy but follow the logic of work processes. A good chef prepares everything in a strict system before starting to cook. For example, just as one proceeds with cutlery on the dining table from outside to inside, cooks place knives and herbs in the order in which they are then used for food preparation. Each knife corresponds to a herb to be chopped.

I am explaining this because it is precisely this thought-out and conscious system of strict order that allows the chefs to concentrate all their skills and creativity later on what is really important: cooking. Instead of looking hectically here and there for the knife while burning the butter in the pan and chasing after sundry items and ingredients, they can flourish freely because they are operating in a cleverly thought-out system of rules. Their attention can be given to the values that they want to express: creativity, self-development and culinary delight. The American author Matthew Crawford, from whom I borrowed this cooking example, speaks of "jigs" (template, precaution), which one sets up and within whose borders one then acts.

What does this have to do with our digital lifestyle? I imagine a world where our digital devices could increasingly serve us with such jigs. They could enable us to not only cook in a concentrated way but also to read and work in a concentrated way. I assume that with the adoption of software agents, we could, for example, tell our virtual butler, "R2D2 I want to work now. Please do not let any messages pass through and shield me from anything that could bother me for the next two hours. Please allow only emergency calls." R2D2 would then seamlessly filter all incoming messages for me, while also informing the people who call me in a very charming way as to why I'm not reachable.

R2D2's program would probably have to run on our smartphones, which the developer community nowadays sees as the central "hub" for our "body area network," i.e., the device we carry on our person that centrally manages all our apps, glasses, smartwatches and of course the AI we use. However, as we already know from experiments, even the mere presence of our smartphones on the table or in the room significantly reduces the quality of our conversation and thinking.[355] It would therefore make sense to develop jigs for our smartphones as access tools to the digital world as well. Precautions or even rituals could allow us to handle the devices more consciously. Symbolic objects and spatial arrangements have always supported our lives and work. For example, we have a garage for our car and a wardrobe for our clothes. We create places for our rites. If it is true what many students have told me, that they have such a close connection to their smartphone that they feel it is like a body part, then it is high time for us to develop rituals and devise objects like sleeping boxes or "phone hotels" for our devices that can help us to be free of them and thereby find some much-needed inner peace.[356] Even smart homes – for which today all sorts of sensors, microphones and video cameras are offered – should be reimagined with this in mind. What ethical dispositions or precautions do we need to respect in order for these homes to develop value qualities that optimize our need for freedom? Do we not fundamentally *need* the offline mode as part of the apparatus of our lives to experience, for example, intimacy, tranquility, privacy and quiet enjoyment, all of which contribute to our freedom?

This assignment we give to our software agents to shield us and create room for thought, or simply to help us get over our addiction with technical help, is not new. For example, iPhones are already equipped with a night function that will not allow calls or messages to be sent or received during the allotted hours of sleep. At some point we forget that we have actually activated such a night function: just like brushing our teeth, it has become such a habit that we don't even think about it anymore. Night functions are time-ordered systems, of which many more are conceivable. For example, we could configure mail and message services to send and receive messages only twice per day. Or, depending on the job profile, companies could give employees an allotted daily time window to process messages so that they can concentrate on the essentials. Some students in an interview study I conducted wished that the Internet could be turned off for everyone at 9:00 p.m. so that people could meet again on the street and develop the neglected values of friendship and recreation.

In addition to ritual places for our digital companions and screen time limits, the next step could be to have freedom functions that are spatially ordered. For example, for students, we could configure digital services so that no messages or phone calls arrive in the library or lecture hall. Or we could make sure that we receive no news in restaurants, sports clubs and the like – generally, any place where community with others ought to take precedence. Public institutions such as universities could even become cell phone-free zones. For good reason, one of the first steps taken by President Macron's French government was to place a ban on cell phones in schools, an initiative now taken up by many governments worldwide.

8.3 Liberty and the Power of AI

Philosophy systematically differentiates between positive and negative freedom. Negative freedom is when there are no obstacles, barriers or limitations, so you can fully do what you want to do.[357] John Locke and Voltaire described this form of freedom as the absence of physical coercion by others or authorities.

Digitalization can have a tremendous impact on how free our societies will be in the future. After all, machines are physically much stronger than we humans. Robots can kill people within seconds.[358] They can lift a thousand times more than their own weight.[359] Even for humanoid robots who have long struggled with the problem of balance, shopping bags weighing 45 kg are light work.[360] The latter run at a mere 14 km/h, moving at the lower range of robot speed. However, systems that are not built with a human form can reach speeds of up to 30 km/h, not to speak of drone technology.[361] Whoever builds and owns such technologies will be able to remotely and forcefully restrict human freedom if needed.

In the past such physical restrictions on freedom were mainly governed by a "state monopoly on force," which is one of the basic principles of the modern rule of law. Since governments that exercise this monopoly cannot survive without the vote of their citizens, they have an inherent interest in maintaining a balance between the granting of liberties and the forcible restriction of freedoms through technology. This is not only true for democratic but also autocratic societies. "If a man is denied the right to live the life he believes in, then he has no choice but to radicalize," Nelson Mandela once said. Perhaps it is this fear of citizen radicalization that gives me hope that at least democratic governments will never be so stupid as to buy humanoid robots from tech companies and use them as police officers against the citizens of their country.

However, I cannot help wondering what would happen if someone other than our elected representatives were to deprive us of our freedom? In the past decades a significant part of the technological power has shifted from governments to businesses (at least in the Western world). A company like Alphabet, most commonly known as Google, owns (at the time of writing) 10 leading robotics companies,[362] two drone manufacturers, its own airport,[363] a Dornier Alpha Fighter Jet (allegedly),[364] autonomous vehicles, communication infrastructure, Google Earth and its high-resolution satellite and street images of all regions of the world, as well as the most powerful AIs that probably exist on earth. Alphabet had sales of around $350 billion in 2024, which is roughly the same size as Iran's or Chile's projected 2025 gross domestic product.[365] Apple, meanwhile, made $391 billion in sales in 2024, which in terms of GDP is about the same as the Czech Republic's GDP. Equipped with such technical and financial resources, tech companies are in a position that even puts warlike confrontation with nation states in the realm of the imaginable.

The scenario that firms might be in military conflict with governments was strangely alien to me when I wrote the first edition of this book in 2019 in German. The idea probably seems outlandish to the majority of employees at such companies. For example, in 2018 many Google employees protested when it emerged that the company had signed a contract with the Pentagon to use Google's AI to fine-tune drones' weapons capabilities.[366] Some employees even left the company. While only 4% signed an in-house petition calling for Google not to involve itself in the production of military technology, the company initially pulled out of its contract with the Pentagon in response to this protest.

Although some components of today's product portfolios could be and are used by high-tech companies for military purposes, such movements among employees show that the bosses of high-tech companies have to at least think carefully about whether and to what extent they cooperate with the military. At this point in time they still have the choice to consciously not engage in this field. If they committed themselves to Value-based Engineering (as described previously) then their IT systems could be built unfit for war by design – or at the very least, they would not allow the atrocities that are reported in military contexts today, such as the use of AI weaponry in Gaza[367] or drone technology in Africa.[368] How tech companies can stay out of war is a matter of their leadership's wisdom and pacifism. They have to weigh that many of their brightest employees might not want to see their work abused for the most aggressive military purposes. That said, the temptation to be militarily involved is hard to circumvent for today's tech-leaders. The American authors P.W. Singer and Emerson Brooking write: "They [the high-tech companies] may want to act like they're not in the business of war, but the business of war long ago came to them."[369]

Against this background, I would recommend to governments to introduce some precautions, such as mandating the openness of tech-companies' research laboratories to include the review of all emerging technologies they build. In the predevelopment stage (or during substantial upgrades) all technologies should undergo rigorous ethical risk analysis or even Value-based Engineering. As described in Chapter 6, the use of such human-centric system design methods in technology development is called ethics by design. In such an ethics-by-design approach, the value of humanity's freedom of action over machines would be granted a special place of protection. It will probably become clear that people should always have the last word in interacting with machines and get fine-granular control over the technology. This means that machines should only be allowed to act autonomously where ethically viable and where people are present: the freedom of the human should always be given precedence over the machine.

8.3.1 Deprivation of Liberty in Small Steps, or Death by a Thousand Cuts

I hope we will never face the fate of being exposed to unlawful or even belligerent machines. But even if this hope is fulfilled, there is another danger to our freedom of action vis-à-vis machines. This danger is much more subtle and omnipresent. I call it Technology Paternalism.[370] Technology Paternalism refers to cases where someone (for example, the state) cares about us and, out of this concern, restricts our freedom. Many car drivers, for instance, are still annoyed that they have to wear seat belts. Occasionally one can observe very funny workarounds in cars, for example, when freedom-loving drivers cut the buckle off the belt and then put it (without the belt) into the locking device in such a way that the safety sensor is satisfied and the annoying beep is silenced. This example shows that the desire for freedom in different people in different contexts can be very varied. Well-intentioned safety technology, whether legally anchored or not, can for some people mean a loss of freedom they are not willing to accept.

The problem is that technology is developed more and more in this paternalistic way, forcing us to adopt supposedly better behavior. And it's not us, the oh-so-fallible people, that decide what is "better" for us, but politicians, lobbyists, companies, innovation teams and engineers. However, these paternal figures do not always have our best interests at heart but rather their own commercial interests. So, a development could be set in motion that means that more and more decisions are involuntarily taken away from us. This is happening, for instance, with increasing frequency in our cars: unnerving speed-limit alarms, attention alarms, lane-keeping alarms, etc., overwhelm drivers and undermine their autonomy. It cannot be ruled out that fridges of the future will tell us that certain foods have not been eaten, or maybe they will refuse to order any food that is unhealthy. Perhaps my coffee maker will only make decaffeinated coffee because my heart is too weak. My car may not start because it measured via odor sensors that I drank a beer.

Such scenarios are not fantastic but technically mature today, and many manufacturers of products that are looking at the possibilities of digitalization in their field are considering them. In technical jargon we speak of the "Internet of Things" or of "ubiquitous computing." When professionals work on ubiquitous computing today, it's all about equipping any objects, nature, infrastructures, and even humans and animals (and of course, cookies!) with barcodes, chips and sensors, and then networking everything – especially to support "better" decisions or increase our presumed "security." Everything should be running according to the wishes of some commercial engineering *paters*.

The weaving of the digital fabric into everyday life was originally thought of positively. The idea is traced back to Xerox pioneer Mark Weiser, who once wrote that "the most fundamental technologies are those that disappear. They weave themselves into the fabric of everyday life until they are indistinguishable from it." Weiser had in mind that one day we would be able to move freely everywhere; for example, in airports, in public transport, at ski resorts, cinemas or concert halls. Nowhere would we have to show barcode tickets or queue up. Since the digitalized infrastructure would recognize us, we could easily go through any gateway or entrance without showing any tickets. For our houses and cars we would not need keys anymore. We could create a world with fewer barriers! Thanks to the digital communication capability of all things, our software agents and robots would do the shopping for us, operate the coffee machine, or facilitate our mobility in the form of autonomous vehicles. The end vision of the Internet of Things was not of systems providing sanctioning feedback but rather quietly going about their business, which is why some in the field also spoke of "calm computing."

Even though all this sounded great in the early 2000s, strangely central to all this networking of things seems to me to be the vision of the "land of milk and honey." Here, roasted pigeons simply fly into people's mouths. In this future phantasmagoria, we are all completely free of everyday activities. Everything that could happen would have already been addressed for us by the *paters*, freeing us to devote ourselves to higher goals. At this point, however, the question arises of whether we really want to be so "free" in this way. The highest value of mankind is life itself, which expresses itself in activity – in free activity – which can unfold outside and beyond the auspices of any kind of patron. It is in truth the many little activities like shopping, making coffee or making appointments that everyone can do which are the lifelong building blocks of our existence. For many, they are life par excellence. Freedom does not only begin when it comes to making major decisions or moves. Instead it determines our existence in the world from second to second. When machines penetrate into this personal space, they not only take from us the freedom to act but also the opportunity to grow and have fun by busying ourselves with the little things in life. In this context, Amartya Sen once coined a term worthwile remembering: “agency achievement” – satisfaction through the possibility of being involved in the result itself.[371]

8.3.2 Frustration with Digital Freedom in Companies

One place where we already find early forms of ubiquitous computing is within companies. For almost four decades now our work structures have been digitally automated. With enterprise resource planning (ER) systems, organizations are increasingly driven by digital workflows. Many production lines are sensor-based. Employee activities are automated wherever possible, recently with more and more AI functionality. The workforce, in contrast, feeds IT systems with information, monitors processes and reacts to key performance indicators that are displayed on computer terminals. The role of employees in companies has changed as a result of these ubiquitous computing environments; they have turned from active agents to passive observers, or what is called “the final decision maker.” This role of being ultimately made responsible for the machine processes sounds positive at first. However, one should seriously consider the implications of this shift in roles against the backdrop of the human desire for freedom and control. What happens if the workflows of IT systems are not completely transparent to the responsible employees? What happens if the IT system makes mistakes? How do people react to a lack of information control?

Watzlawick’s experiments, which I presented in Chapter 4, give an idea of how people are trying to regain control over the IT systems they (usually) do not understand. It is not merely that humans are physically uncomfortable with a lack of control – they also go into autocomplete mode and invent theories about how the IT system is likely to work. This path of half-truths is often a thorny one because false explanations to cover up computer errors or gain some plausibility for something incomprehensible are destined to fail at some point. Many employees struggle with not being able to fully understand an important part of their job. I have seen firsthand how desperate people become over IT systems they do not understand but are still made respsonsible for. They lose sleep, develop abdominal pain or rashes. They are experiencing what American sociologist Robert Karasek describes as burnout because they have no control over the tasks assigned to them. Across Europe 66% of people have either had burnout or feel on the edge of burnout, a steady increase from the previous high of 40% in 2010.[372] And there is decidedly little thought about the connection between the progressive digitalization of corporate processes and burnout quotas.

Burnout is a tough fate for many employees, but equally awful is the fact that many people no longer enjoy their highly automated jobs in which they spend

50% of their time awake.[373] Even if information control is satisfactory, IT systems limit employees' hands-on activities and active decision-making. The spectrum of decision-making options, i.e., the naturally available degrees of freedom and the ability to improvise and adapt to real conditions, narrows substantially what humans can do.

What happens when employees' degrees of freedom are reduced through automated processes is clearly demonstrated in today's airport check-in areas. Prior to automating the procedures, stewards and stewardesses had the unspoken freedom to decide whether or not to let a passenger pass through with her excess baggage. These were the trivial, short and beautiful moments where the check-in staff had the power either to be strict with passengers or do them a huge favor. Values such as goodwill and generosity on the part of the stewards would have fostered values such as gratitude and joy on the part of the passengers. Those days are over. Computer check-in systems fully automate such decisions. At today's check-in machines the few left-over employees can only listen to the grumblings of passengers regarding additional payments or watch them haphazardly rearrange things from one suitcase to the other until the implacable digital scale finally lights up green to signal that they are allowed to pass. The hustling social system at check-in areas has been abolished. Airport halls have become void and anonymous places, just as most offices have become nondescript desk-sharing spaces without soul.

Looking at these developments, it is not surprising that 79% of employees are not really motivated in their jobs anymore.[374] What free person would like to work voluntarily within such structures, from which the lifeblood of human interaction, decision-making, mutual help, etc., has been completely sucked away?

8.3.3 Nudging and Manipulation

The airline example shows an economic dynamic that plays a central role in the creation of digital structures: profit maximization. If every passenger grudgingly pays their excess baggage, the airline earns more money, and if they do not, the airline saves kerosene because less luggage means a lighter plane and greater fuel economy. Moreover, by fully automating check-ins staff can be cut, which reduces labor costs. All this makes business sense. It cannot be ruled out that such economic dynamics will also play a bigger role in the automation of our everyday lives in the future. Technology Paternalism can certainly pay off. Health insurers would be happy to subsidize refrigerators that only contain the right nutrition and would pay premiums to obese policyholders if they allowed themselves to be technologically supported in their eating habits.[375] A real-world example is the

Go365 Health App, which was approved by the public insurance company of West Virginia (USA) for all public servants. It counts all steps and registers all whereabouts, as well as monitoring sleep cycles, sugar levels, diets, and smoking and drinking habits. The insurance company said that the use of the app was voluntary; however, it rewarded users with bonus points that had to be achieved and penalized those who failed to earn these bonus points with hefty extra insurance payments of up to $1,500.[376] This example shows that it is not only smart objects such as cars or refrigerators that directly limit our freedom of action through warning signals or actuators but also the data that is collected by such systems. For example, someone whose health insurance depends on his eating habits and who has an appropriately "intelligent" refrigerator will certainly feel limited in his freedom to eat anything. Someone who has the kind of car insurance whose monthly rate depends on driving style will probably drive more cautiously as well. Each time, the deal between us humans and the commercial patrons of our society is: "prove it digitally and you can pay less."

Now, we could argue that many people would indeed be well advised to eat more healthily or drive more sensibly. It is true that people constantly do things that are not good for them. Our questionable image of mankind can find a lot of affirming examples. This leads many meritocratically minded leaders to see direct and indirect technical controls as salvation for society. They can finally "nudge" the lazy, irrational human race into more rational behavior. To "nudge" is the new magic word of the elites.[377] The question is only how it is done. A lot depends, in fact, on how friendly or funny the apps are that encourage freedom restrictions. A display on my fridge could, for example, show me my own (terrible) body figure as a cartoon and playfully visualize my fat-rolls when I reach for the chocolate ice cream for the umpteenth time. Such visuals would give me pause for thought and perhaps help incentivize the values of beauty, health, attractiveness and self-love that I also cherish. The example shows that nudging can be done in a good way. However, I support this kind of impetus only under three harsh conditions that the creators of today's nudging world would not care for at all – rules that once more underline the importance of the art of omission: First, nudging practices should be completely voluntary. There should not be any proactive kickoff settings in technical devices that preimpose them. Instead, people should be playfully asked what they want to be reminded of in terms of good intentions when setting up their equipment. In the interests of my own value preferences, I would like to be able to tailor my levels of personal freedom in dealing with the devices myself.

Second, the value of my privacy should not be undermined by the digitalized things around me. I do not want my health insurance to know about my private lifestyle and give it the opportunity to increase my monthly installments. This is precisely what happens if refrigerators and pedometers "send home" user data to an ecosystem of companies and insurances.

Thirdly and finally, this responsible design of my free space must not be influenced by financial incentives. The car insurance premium, the health insurance premium or the price reduction of goods should never be linked to my "good" or "bad" behavior. Why not? Wasn't WeightWatchers so successful precisely because customers had to pay if they couldn't cut the mustard? Clearly, financial incentives are very powerful in controlling our behavior. But money and freedom should never be balanced against each other because too many people have far too little money and are always the losers in such a game. Too many households are already living in a situation where they have to count every single penny. If the Internet of Things were used to seduce these poorer and middle-class households, exchanging their freedom for cash rewards, that would result in a dramatic indirect loss of freedom for many citizens. Fifteen dollars for sensible driving, $15 for healthy food, $5 for 10,000 steps a day – at every turn you could save money in exchange for small losses of freedom. After all, a penny saved is a penny earned. The end result is a society that exchanges its freedom for monetary rewards and in which only the rich are free because they are the only ones who can afford not to care about any additional payments. The outcome would be a dystopian division of society into the technologically oppressed and the free.

9 Personal Progress in the Digital Age

At the beginning of this book I suggested that we achieve progress when the actions of each individual are based on values. Everyone contributes to the creation of value in society as a whole through their good actions. The curve in Fig. 2.2, accelerates upward. In this scenario, we would use our powerful technologies for the right purposes.

Such a valuable path is not easy to tread, though. Our modern model-based thinking, the idea that something new is by definition already good, the will to bulldoze existing values, the naive confidence in the perfection of AI, our lack of trust in humankind, our purely profit-driven business models and our lack of understanding of the world of values – all these are obstacles to implementing the beautiful, almost paradisiac progress that my students have outlined for FoodIS.

Just because challenges lie in wait does not mean, though, that we should not set off on that journey. “Courage–that is quite certain–is the most necessary of all human qualities for happiness,” said the pedagogue Johann Heinrich Pestalozzi (1746–1827). We need courage for the future. Well then, if the goal is clear and the courage is summoned, the question arises of which way can and should be taken. What can we do to concretely contribute to such a humane and trustworthy progress as individuals, in our diverse roles as private persons, as employees, as customers, as entrepreneurs or even investors? How can we personally progress and thus contribute to the overall progress of a digitally permeated society? In Chapter 6, I already suggested the path available to us, which roughly consists of three steps. The first step involves being attentive to values. The second requires building a deep understanding of prioritized values. The third and final step is to develop valuable habits. Each of us can take these three steps in our daily use of the digital. But not only that: these three steps are also independent of the digital, and guide us on the paths that the great thinkers have shown us since ancient times, paths that lead to a good life full of individual progress.

9.1 The First Step: Being Attentive to Values

The first step to true progress is reflection on our own value goals, just as my students exemplified the potential value goals of FoodIS. No matter how far we have journeyed, we should always take the time to reevaluate how good things really are at our current port: If we have the desire to continue to progress positively

 | https://doi.org/10.1515/9783111677736-009

and sail to an even further harbour, then we should truly know *why* and *where* we want to go. The *why*-question is seldom considered in digitalization. As already stated, we believe everything new is automatically good. But this is not true. It is essential to ask at each new automation of our lives what positive values we actually gain from it. What are the positive value goals we desire for ourselves, our family, friends, companies and corporations, universities, customers or employees, beyond this simple credo that new is good? Or is there potential that by investing in IT and AI we actually promote more negative values than positive ones? After reading this book it should be clear to you, dear reader, that I am not speaking about monetary return on investment here. Successful entrepreneurs like Steve Jobs had an answer to the why-question beyond money. He knew that true progress is not only about unilateral profit maximization. Rather, it is about the unfolding of a multitude of human and social value potentials that our societies are in need of or cherish.

9.1.1 Small Steps to Being More Attentive to Values

We do not automatically run through the world with an attentiveness to values. It is something to work on and to pay attention to. Being attentive to what you are doing is what Buddhists refer to when they speak of "mindfulness." Being attentive does not mean that we should be judging our surroundings at all times with our own preconfigured value lists. It is rather that value-attentiveness requires us to take a step back and resist the natural inclination to judge everything immediately. It's about really seeing the world around us, being alert, empathetic, and listening. Attention is necessary to distinguish truly appealing positive values from repulsive negative ones. What is really going on in my family, in my circle of friends, in my company or in my market? This is a question we must first ask ourselves, immersing ourselves in visible and invisible reality. So, instead of automatic recourse to AI advice, future-studies, analyst reports or analyzing models and statistics, it is important to pay homage to our personal notions of desirable values and futures. We must engage in regular discourse with the world.

In the original preface to this book published in the 2019 German edition I described my beginnings in the IT world at Apenwaze Systems. Back then, as I said, all future predictions were based on the fact that the operating system Symbian would conquer the mobile Internet world. For this reason, I conducted market research in actual stores and tried to point my company to real trends on the ground. On a business level, that is exactly what I recommend to today's companies and politicians. See what is *real*! Does AI deliver the return-on-investment it promises? Are promises from LLM providers of true super intelligences warranted when the marginal performance rates of models are slowing down and the value-alignment problems are not solved? Can we afford the energy expenses to really scale AI for everyone? Are nanochip supply chains not far too fragile for companies to rely on them in the near future at vast scale? Are human employees and contractors not ultimately more reliant and even cheaper than AI solutions that must be serviced in the long term? All these questions should be on decision-makers' tables when judging the economic value proposition of AI.

On a private level, you should do exactly the same thing. Rather than reading in a random report that a new scientific study has shown two hours of video games a day for children to be unproblematic, parents should ask themselves the question, What are smartphones and AI assistants doing to my child? How do these devices and services matter in my relationship with my child and influence our family?[378] How does my private world feel? Which values do I give space to, and which ones am I currently neglecting too much? Politicians should ask themselves the same question and spend more time in their local constituencies talking to ordinary people. Business owners should do more "management by walking around" instead of visiting future conferences, reading e-mails or talking about the blessings of automation at the Rotary Club.

In the course of such value-seeking practices, the three philosophical questions outlined in Chapter 3 may be helpful: Which virtues and vices or personal qualities has the digital helped to develop in our social environment and in ourselves? Which advantages and disadvantages should be considered? What are my own value priorities that I consider to be universally important? At this point, it should be remembered that the value questions should not be primarily formulated negatively. Rather, we should also consider the positive values we wish to give more space to. For example; for years I have been wondering how to deal with the flood of information that reaches me as a knowledge worker. I have already explained to what extent we are more and more locked into e-mail, messages, social networks, etc., leaving us with barely any space to breathe. Many of us would, therefore, like to live more freely and more independently in the future. Freedom is a key value of our time; but already in the year 2003 when I worked for Apenwaze, the wish to regain my freedom was hard to fulfill. I faced the problem that already at that time most of the communication in the company was messenger- and e-mail-based rather than personal. I was probably inundated by 200+ instant messages a day announcing every colleague's arrival (log-in activity). This senseless overcommunication stretched me enormously. Values that are most dear to me – my freedom, my health and my productivity as a knowledge-worker – suffered to the point where I could only work pain-free by lying down on my kitchen floor. I was only 29 years old at that time. In such a situation, it was not hard for me to recognize the negative values that overtook me: not only ailing health, dissatisfaction, overwork and frustration but also superficiality and lack of productivity. I wondered what I could do to give my freedom and health – as well as the quality of my work – more room again. So I decided to turn off the chat and rejected all instant messages from colleagues. I put filters on my e-mail account that automatically archived all CC and BCC mails in a folder called "impersonal." All news feeds went to subdirectories titled "to read," (and which I quite honestly seldom opened). As a result, the number of mails in my main inbox shrank substantially within weeks. I regained my health and became much happier. The quality of my market studies were great; and yet one day, my boss confronted me, full of reproach, with the question of why I was not visible in the chat. It was an unpleasant moment. Accessibility was a value that was put on a pedestal at Apenwaze. Everyone in the company was available at all times, even though many certainly suffered like me. I explained the situation to him and emphasized that the quality of my work went above and beyond. I would not be able to keep it up if I had to expose myself to a constant flood of messages. To defend my attitude required courage. In the end, the chat stayed off. My boss had no choice but to come to terms with it. But he still accepted it since the knowledge I produced as well as my health and productivity were also far more important values to him than my instant reachability.

As we stand at the crossroads of true progress, it is essential that each individual recognizes the right mix of values for him or her and has the courage to take decisive steps to defend or regain them.

9.1.2 Bringing Personal Value Priorities to the Community

In every situation there are many positive and negative values in play. We are surrounded by numerous value-bearers at all times: the technology itself and its many settings, which influence our well-being, our privacy, our freedoms, etc. Diverse relationships, symbols, and business models: all carry values that have the potential to speak to us and to influence us. In order to respond appropriately to this immensely rich invisible value world around us, we must constantly set priorities and reflect on the *rank* of values at our fingertips

The philosopher Immanuel Kant coined the term "personal maxim" in formulating the categorical imperative. Maxims are principles of personal will and action that are considered so important that they are believed to qualify as a universal rule. For example, dignity is so important to most people and at all times that we must always be careful to not undermine it, whether in our own doings, through our business models or by using and configuring technology in the wrong way. Regular reflections on our personal maxims in the sense that Kant intended might certainly support wiser actions, also when it comes to dealing with AI. It strengthens our character to be aware of our maxims and to reflect from time to time on whether they have a claim to be valid for other people and contexts as well. Think of the value of dignity in the face of AI. Isn't it below our own dignity to pretend that an AI-written text was authored by ourselves? Don't we lose face and self-respect when we know less and less because we delegate all cognitive tasks to AI?

I would argue that my own value priority at Apenwaze at the time and the courage to defend my freedom reflected my personal maxims at work. And I wish that others would similarly prioritize their own dignity, freedom and mental health over constant digital accessibility and responsiveness. But that does not mean – and here wisdom comes into play – that digital accessibility is always less important than other values. There are situations, such as in emergency cases, where accessibility and responsiveness are absolutely necessary and indeed more important than anything else.

How to set value priorities correctly thus has much to do with one's context. It is influenced by personal experiences, one's upbringing as well as a good dose of self-knowledge. My friend Doris demands permanent reachability, responsiveness and accessibility from her children and calls them and tracks them with their smartphones constantly. Those children, once adults, will certainly prioritize the value of accessibility differently than I did at Apenwaze. Perhaps they will find accessibility and responsiveness much more natural and important than freedom or work productivity because their mother taught them so. Maybe, on the other hand, they will reject accessibility violently because their mother has forced it upon them. Who is right? Myself with my tic for freedom and productivity? Or Doris with her insistence on accessibility, readability and control?

A meaningful life is not per se about who is right but rather about recognizing the real value priorities existing within the life context of our community. If the family of Doris, where responsibility for each other is very important, were in distress, accessibility and responsiveness are value qualities that would here foster unity, trust and conviviality. Accessibility, then, is not an end in itself, but really strengthens higher intrinsic ends. However, this will not always be the case. I find it fascinating that, as I previously mentioned, the exhortation "Know thyself!" was inscribed on the gate at the oracle of Delphi. Because perhaps it is not the plight of the family that drives Doris to bring up her children to be so accessible but rather her own inner emptiness, her boredom or her obsessive need to control? It is self-knowledge in the wake of self-reflection that allows us to recognize why we use technology in one way or another and unconsciously prioritize values incorrectly or correctly. This self-knowledge is crucial to a good life with technology, as well as to leadership in digitalized organizations.

Perhaps the questioning of value priorities could lead entrepreneurs, investors and managers to ask why they so often actually prioritize the values of profit and efficiency over higher values like dignity or true progress. What makes them value, think and act in this way? And which values are subsequently promoted or destroyed in their own corporate community as a result of false value priorities? Can they stand up for the values they promote in the long term?

9.1.3 The Right Role Models for Progress

Good role models might help us to find our value priorities. As a teenager, I encountered through Feuchtwanger's novel *The Foxes in the Vineyard* the great American scholar and statesman Benjamin Franklin (1706–1790), who kept a list of personal value priorities. This list contained 13 virtues which he had set as his personal priority: abstinence, silence, order, resolution, frugality, industry, sincerity, justice, moderation, cleanliness, chastity, tranquility and humility. I cannot say with certainty whether I could live such a disciplined and long virtue list, but I consider his intentions quite exemplary. He must have been a man in the mold of Kant's "autonomous" thinker, who had not only set himself these value priorities as his goal but was also courageous and disciplined enough to live accordingly. But are we all so strong?

When Kant formulated the idea of personal maxims, he was hopeful that we humans would all be autonomous and strong enough to live up to them. Theoretically, if we all go through an intense process of self-reflection, we will come up with a similar list. And certainly, there are role models that seem to do justice to such a positive image of humanity. Perhaps the Dalai Lama or Pope Francis could be attributed such exemplary status. They are role models who have proved they have the courage to develop and exemplify their own maxims. At the same time, however, they are also wise enough to weigh up phenomena in a situation-specific way. That is what makes them such strong and authentic leaders. Most of us will also know some wise old people or even a young person in our circle of friends and family who are exemplary in the same way. However, as behavioral psychology and history shows, people are not always as independent and strong as these leaders or the brightest of our friends. That said, it is enormously important to reflect on which role models help us orient ourselves in our own progress towards virtue.

The importance of role models for our own development has been recognized in almost all major cultures. Aristotle called a man of exemplary wisdom a *phro-*

nimos if he was able to recognize the good – that is, the values – for himself and others. In Confucianism, the word *junzi* stands for such a role model. Junzi is also translated as "aristocrat," in the sense not of a man of aristocratic blood but rather a naturally noble man. The term "aristocrat" is in fact composed of the word *aristos*, meaning in English "virtuous" or "excellent," and the word *kratos*, which means "power" or "strength." Following this etymological logic, good role models are therefore about the marriage of virtue and power. Wikipedia quotes a nice description of the junzi, whose example was supposed to be followed by educated people in China who intended to hold public office: "*Junzi* have many characteristics. A *junzi* does not compromise his virtues even in poverty; a *junzi's* actions speak louder than his words; a *junzi* is loyal, obedient and knowledgeable. A *junzi* disciplines himself. According to Mencius, *ren* is the core virtue of a *junzi*."[379] (Ren stands for "benevolence" and humanism).

Perhaps it could also be beneficial for today's entrepreneurs, investors, politicians and managers to reflect afresh on the question of whether they themselves are role models for the many people they lead and whose destinies they influence. Do they have enough self-knowledge to deal modestly and virtuously with others? Have they studied sufficiently the values of the social systems that they run and the value potential of the products and technologies they bring to market? Are they "aristocrats?" These are the questions I would like them to reflect on while they are struggling for wise leadership.

9.1.4 Self-Awareness and the Importance of Teleology

Many believe that they are doing the right thing if they simply mimic others and accept the values or virtues of their own environment. While that is true to a degree, with true self-reflection on one's own progress it is not enough to just mimic others. Instead, it is also a question of what values we feel inwardly drawn to and what incontestably repels us. This is the question of the order of one's heart. How must the value-bearers around us, the people, the relationships, the things be arranged for us to be at ease? And what do those who encounter us in real life say about us? Are they perceiving us as valuable for them? What role do they see in us ? And can we learn to love this role? In the subtle observation of a person's own situation and preferences lies the seed of what Aristotle called *telos*.

Telos is the potential within each one of us from birth onwards to develop certain goals or purposes in life, a unique stamp of one's own character. The telos shows itself in the repository of talents whose early tendencies our mothers already noticed in us when we were still lying in the cradle. Reflecting on our own behavior in community or our own preferences, we can recognize parts of this personal telos. If we look closely, we can see that we tend to always prefer certain things, that we are attracted to certain values and types of people, that we always make similar decisions, or take similar roles in groups. All this is an expression of our own telos. Research in motivation psychology tells us more about telos today: There are people who have a very strong urge for power from an early age onwards, while others tend to seek friendship and harmony. For yet others, a perception of achievement and deep truth is vital.[380] Some are born with a permanent smile on their face: they see a world full of positive values. Others, meanwhile, see a world full of negative values and easily become cynics. Part of self-awareness, or self-consciousness, is ideally that we recognize our own telos and delve deeper, so that we understand the value potentials that are given to us.

To be able to live our own telos is the original principle of the English garden I introduced in Chapter 5. The English garden principle does not try to shape man and turn him into a trimmed boxwood tree. The English garden looks at natural value systems and the pursuit of everything that lives, and skillfully integrates these potentials into a culture. Living our own given telos means trusting in the good and in the uniqueness of everything that lives in our garden; humane progress, then, means to skillfully and carefully combine these potentials into a meaningful unity.

Finding and pursuing our telos is not easy in the modern digital culture of progress. As described above, our current scientific thinking is focused on theoretic models and a notion of control over nature. It wants to proactively shape our humanity instead of seeing each individual naturally unfold. The modern digital culture of progress has no understanding of the singularitas of individual life or its telos. Even Goethe recognized this Zeitgeist, which began even before his time. He wrote, "When the teleological way of explanation was banned, understanding was taken away from Nature; there was not the courage to ascribe reason to her, and she finally remained lying spiritless."[381]

9.2 The Second Step: A Conceptual Understanding of Values

After going through the process of self-reflection, we can critically recognize certain value priorities we have as a person or as an entrepreneur. We might want to strengthen them and follow our own telos more rigorously. But how can we do this?

The true step required is to forge a deeper understanding of our own value priorities. In an earlier chapter, I demonstrated how complex an elaboration of the values of privacy, knowledge and freedom can be. If a company like FoodIS wants to nurture a value such as loyalty among its bike couriers and customers, then it first must understand what loyalty actually *is* and how it could be encouraged within its own corporate structure. If community and friendship were to be promoted amongst couriers and clients, what do these values actually stand for? We constantly talk about all these values as if we knew exactly what they mean. But do we, really?

A decade ago, I went through a thinking experiment that from today's perspective seems totally awkward to me, but is still symbolic of how we often think as children of our time. I wanted to be more aware of my friendships. Seeking to better understand the true meaning of friendship, I decided to do exactly what our science world does today: I built a model to characterize, measure and benchmark my friendships. I designed a list of criteria that matter to me, such as whether someone is responsive, whether he or she likes me a lot on Facebook, or if I see someone regularly. Then I tried to rank all my friends and acquaintances using these criteria. But somehow, this whole exercise did not fit. I started to attribute weights to certain criteria, which is what you usually do in model development. But time and again, people in the ranking whom I intuitively knew were not my best friends leapt forward.

I learned two things from this exercise: First, I had no idea what friendship really is. I knew the value, whose name I had often lightly tossed around in speech, actually only intuitively, but not in its multi-layered depth. I had never read the numerous books written about it since Aristotle. I have had so much stuff to learn in the course of my life, and yet, theoretically, I knew nothing of the ethical value of friendship, which is so important to me. That is why I tried to decode it in the first place. But it just didn't work, because – and this was my second wisdom gained – ultimately, values cannot be dissected and turned into a measurable model. You simply cannot break a value down into pieces and then mathematically sum up how much you have of it. Values like friendship are phenomena that are constantly in the flow of time. They are newly constituted by life, moment by moment, in every situation. When values unfold, we recognize them; but if we want to apprehend them, then they suddenly seem to have disappeared elsewhere, evanescent. Nonetheless, generations of philosophers have been concerned with value phenomena and have tried to at least roughly describe them. This is how Aristotle came to write about the basics of friendship; about what in Greek is called *philia*. In Greek thinking, philia is not limited to the immediate circle of personal friends, but can refer to far more people we encounter, whether in academic life, in politics, on vacation or in virtual games. It is possible to intuitively feel this philia with someone, and it is essential to a happy life. However, there are three kinds of friendship that Aristotle distinguished: utility-based friendships, pleasure-based friendships and true friendship, that is, virtuous friendships.

Such a deeper, even philosophical understanding of values is fundamental in the digital age if we want to design the digital and AI specifically as ethically as possible. If we do not understand what values stand for, how can we make them priorities? We would be like boatmen who know their desired goal but do not have the ability and the knowledge to cross the seas. Only if we know what friendship is can we also find the measure of what level of digital exchange with friends is appropriate – and thence, where the digital should be switched off, or is not the right medium.

Companies, too, have to grapple with the full depth of values when they realize that these are fundamental to their business model. So I would assume that for social networks like Facebook, the value of friendship should play a role. If such platforms seriously grappled with this value, it would be possible for them to systematically support friendships, even high forms of friendship. For example, the provision of a "rich" communication interface via video call would be more conducive to friendships than a mere text chat. When friends talk openly, such exchanges should also respect the privacy of people. How can I tell my friend something openly, perhaps discuss a weakness with him, if I have to assume that Facebook records and resells everything that we talk about? It cannot be that one maintains friendships on a platform and then an algorithm decides, based on obscure criteria, which few of my friends see my news at all. Or whether friendship is replaced outright by parasocial relationships with influencers.[382] Friends have to decide among themselves about whether or not they see something of the other person, and this also applies online, otherwise the criterion of reciprocity is not met. In short, the profound knowledge of values is indispensable for the design of technological products and services.

9.3 The Third Step: Habits, Technology and Politics

When you are ready to recognize and more deeply understand value priorities for yourself or your company, you need to put them into practice. You must strive to develop the prerequisite conditions in which values can unfold.

9.3.1 The Importance of Moderation

In implementing value priorities, we are naturally struck by the virtue of moderation. I have already described how important it is to learn the art of omission so that the right balance of values can be achieved. If we overdo it with our dignity, our modesty or our generosity, then this is not the right middle ground.

To return here to the example of accessibility, freedom and work productivity, it was not easy for me to reduce my accessibility for colleagues and my boss at Apenwaze. I wanted to improve my work productivity, restore my health and happiness, but also still be a good and responsive colleague. Permanent accessibility did not, however, fit with these goals. The word "permanent" is important at this point because, as I explained previously, accessibility per se is not bad. Accessibility is a value that is important to any kind of community. It is a foundation for the functioning of social systems, especially in companies. So I had to find the golden mean of being reachable just to the necessary degree that my colleagues and bosses would still be happy with me while at the same time affording myself the right conditions for undisturbed good work.

The golden mean is a centerpiece of ancient thought about the "advanced" character. It is at the center of the art of omission. According to Aristotle, a wise person has the gift of finding the right measure in every situation and of avoiding extremes. Let us once more take as an example e-mail etiquette and habits. An empirical study showed that e-mails should only be checked three times a day to maintain personal well-being.[383] Based on this knowledge, I try as far as is possible to align my behaviors with this measure. However, if someone from my team needs my help or has an important deadline looming, I abandon this requirement and go online immediately. In a time-limited situation where my help is urgently needed, a pragmatic response means making an exception to the rule.

It is particularly difficult to find a moderate center and to omit activities when our environment tends to extremes. At Apenwaze, excessive overcommunication was the prevailing culture at that time, which led to permanent interruptions. In Chapter 2, I described how the habit of interrupting us humans leads, in turn, to self-interruption. We take these interruptions into our private lives because on any given working day we spend a good 50% of our waking hours with our employers – and once we have adopted a habit at the professional level it often bleeds into the sacred private sphere.[384] So, I have been working for years on the art of reducing my e-mail traffic. By my estimates, it has been 10 years now since I made this resolution; since then I've deliberately worked with only modest accessibility so as not to lose my freedom. I try to make a habit or a practice of giving in only three times a day. These efforts cost me strength and discipline. Getting away from alcohol and cigarettes is probably not any easier, once you've become addicted!

9.3.2 The Importance of Rhythm and Rituals

In fact, a discipline that is not always easy to maintain requires us to develop habits in order to live up to our own value priorities. Once we feel that the new habit is beginning to consolidate, the "virtuous" life itself can become a source of joy. Athletes know this feeling only too well. Initially, the conditioning training to get into peak shape is very tedious. But then there comes a point where you begin to *need* it to be satisfied. When our inner master begins to awaken, we rejoice in the power to give form to our life. It is not without reason that the Greek and Latin word for virtue (*virtus*, *aretē*) derives from the same root as "force."

One way to support and cultivate our strength is to pay more attention to rituals and rhythms. For example, there are people who have their day structured perfectly and do the same thing every evening at a specific time, such as checking digital information in a very specific room at a specific table. In Chapter 5, I explained how structured behavior enables us to more easily act in a fair and good manner. Structures and rhythmic rituals are difficult to implement when it comes to the digital in our networked world because everything and everyone wants to access us at the same time. But this is exactly what we have to nip in the bud. In Chapter 7, for example, I described the idea of literally putting cell phones to sleep in a separate place. These are all just ideas: ultimately, it falls on each and every one of us to come up with a working strategy.

9.3.3 Technical and Political Measures to Foster Values in the Digital Age

The fact that today I manage my e-mail load in the way I do is also in part due to functionalities available in modern e-mail programs. For years, Microsoft has invested considerable amounts of time and money into researching and developing attention-saving IT systems.[385] For example, most e-mail programs today allow us to filter e-mails, mark junk advertisements and set rules that determine which messages are displayed. Likewise, any modern smartphone today allows us to turn off interrupts and prompts in the form of ringtones, vibrations and banners. This closes the circle on the value-based design of technology, as I described in detail in Chapter 5. If such technical functionality did not exist, I too would be helplessly exposed to permanent interruptions.

So, beyond discipline, rhythm and courage, we need technical adjustment options that help us implement our value priorities in this digitally upgraded world. In computer science, we refer to "ethics by design" when it comes to providing such value-oriented settings or ethically balanced defaults. "Privacy by design," "safety by design," and "security by design" have been the termini brought forth by the early ethical movement among engineers. But we are still at the very beginnings of this journey, as I have described in Chapter 5. I have shown how many relevant values we encounter in different contexts at all times, and that every thing or place is a value-bearer. This also holds for our technical devices, with all their many functions. Therefore, in my view, the systematic and comprehensive work on value-based technology with Value-based Engineering must become a hygiene factor in all technical developments. When pursuing Value-based Engineering, companies weigh all the values and impacts of a digitalization strategy with relevant stakeholders and decide in an anticipatory and participatory way what technical and organizational action is needed to create positive values and prevent value destruction.

In this book, I have addressed many suggestions for such a form of Value-based Engineering. I have shown that we can design AI systems and social media that do not make addicts of us, that do not encourage us in grandiosity and that do not entangle us in a web. All of this is possible if we gain a sober assessment of the nature of the digital medium – its limitations, strengths and weaknesses. I have hinted that Value-based Engineering may also at times require that we do not digitalize, perhaps even that we re-manualize certain processes; especially those where human singularitas is important or where machine errors can cause too much harm to individual fates.

Importantly, the political forces of our time must recognize this need for Value-based Engineering and learn to grasp the truly limited nature of the digital fabric. Today, many politicians are entangled in a dense network of lobbyists and experts in whose financial interests it is to keep politicians suspended in the hazy cloud of digital storytelling. Unaware of the alternatives and true design options, government officials march en masse to AI conferences and believe that they have to invest in tech-giants' AI solutions as quickly as possible. In truth, they can only support *true* humane progress if they prescribe value-based technology design and make ethics by design a legal requirement. Such a step would be a kick in the teeth to today's digital business models with their short development cycles. So to roll out Value-based Engineering, and a potential certification for it, requires courage on the part of our political forces.

Particularly important for Value-based Engineering is the art of omission: if we deliberately reduced accessibility in our networks and only got digital news at certain, rhythmically predictable times; if we were officially allowed to get rid of the current push model in communication; if our attention were not proactively consumed by third parties (such as advertising); if tech companies were stopped from configuring robots and AI in a human-like way; and if we said goodbye to the idea that machines could think *for* us. If only we could become capable of such profound measures of omission, then we would have a chance to put the digital economy on a humane path. I wish I did not have to speak in such subjunctive terms here. The list of ideas sounds improbable, even to me. It is a wish. And it is only the beginning. But I'm sure we're at a point in the history of our civilization where we need to rethink. A gigantic reorientation of our behavior lies before us should we wish to pursue value-based progress. Should we choose to do so, we must break away from purely economic rationality and turn to value-oriented thinking. This can be achieved only by taking all those steps described here, and doing so with courage and discipline. As Goethe, in his wisdom, once remarked: "He for whom life is a game can never grasp its worth; he who has no self-command will forever be a serf."

References

Abagis, Tessa. “How Brain Stimulation Can Boost Memory if Paired with Learning.” *Aeon*, March 21, 2018. https://aeon.co/ideas/how-brain-stimulation-can-boost-memory-if-paired-with-learning.

Aboujaoude, Elias. *Virtually You: The Dangerous Powers of the e-Personality*. W.W. Norton, 2012.

Abramson, Lyn Yvonne, Seligman, Martin E. P., and Teasdale, John D. Learned Helplessness in Humans. *Journal of Abnormal Psychology* 87, no. 1 (1978): 49–74.

Aeon. “Aeon | a World of Ideas.” Accessed December 4, 2025. https://aeon.co/index.

“AI Principles Overview.” Accessed December 4, 2025. https://oecd.ai/en/principles.

Aiswarya, Balachandar and Ramasundaram, Gurusamy. Online Food Delivery Technology and Family Food Traditions: A Sociocultural Perspective. *Journal of Information Technology Teaching Cases* 14, no. 1 (2024): 68–73. https://doi.org/10.1177/20438869221150826.

Albanese, Andrew and Milliot, Jim. “Self-Publishing’s Output and Influence Continue to Grow.” PublishersWeekly.Com, November 8, 2024. https://www.publishersweekly.com/pw/by-topic/industry-news/publisher-news/article/96468-self-publishing-s-output-and-infuence-continue-to-grow.html.

Allred, Ryan J. and Crowley, John P. The ‘Mere Presence’ Hypothesis: Investigating the Nonverbal Effects of Cell-Phone Presence on Conversation Satisfaction. *Communication Studies* 68, no. 1 (2017): 22–36. https://doi.org/10.1080/10510974.2016.1241292.

Alter, Adam. “How Technology Gets Us Hooked.” Games. *The Guardian*, February 28, 2017. https://www.theguardian.com/technology/2017/feb/28/how-technology-gets-us-hooked.

Altman, Irwin. *The Environment and Social Behavior: Privacy, Personal Space, Territory, Crowding*. Brooks/Cole, 1975.

“Amazon Turk.” n.d. Accessed September 10, 2025. https://www.mturk.com/.

Anders, Günther. *The Obsolescence of the Human*. Translated by Christopher John Müller and Christian Dries. University of Minnesota Press, 2025.

Anderson, Michael, and Anderson, Susan Leigh eds. *Machine Ethics*. Cambridge University Press, 2011. https://doi.org/10.1017/CBO9780511978036.

Andrijevic, David, Vrselja, Zvonimir, Lysyy, Taras, et al. Cellular Recovery after Prolonged Warm Ischaemia of the Whole Body. *Nature* 608, no. 7922 (2022): 405–412. https://doi.org/10.1038/s41586-022-05016-1.

Annisette, Logan E. and Lafreniere, Kathryn D. Social Media, Texting, and Personality: A Test of the Shallowing Hypothesis. *Personality and Individual Differences* 115, (September 2017): 154–158. https://doi.org/10.1016/j.paid.2016.02.043.

Anscombe, Gertrude Elizabeth Margaret. Modern Moral Philosophy. *Philosophy* 33, no. 124 (1958): 1–19.

Anthony, Andrew. “‘Eugenics on Steroids’: The Toxic and Contested Legacy of Oxford’s Future of Humanity Institute.” Technology. *The Guardian*, April 28, 2024. https://www.theguardian.com/technology/2024/apr/28/nick-bostrom-controversial-future-of-humanity-institute-closure-longtermism-affective-altruism.

Aristotle. *Aristotle’s Nicomachean Ethics*. Translated by Robert C Bartlett and Susan D. Collins. University of Chicago Press, 2011. http://archive.org/details/aristotlesnicoma0000aris.

Arndt, Bettina. “What Gives a First Online Date Every Chance of Turning into a Second?” The Sydney Morning Herald, November 16, 2016. https://www.smh.com.au/lifestyle/life-and-relationships/what-gives-a-first-online-date-every-chance-of-turning-into-a-second-20161116-gsqg12.html.

 | https://doi.org/10.1515/9783111677736-010

Aronson, Polina, and Duportail, Judith. "The Quantified Heart." *Aeon*, July 12, 2018. https://aeon.co/essays/can-emotion-regulating-tech-translate-across-cultures.

Asimov, Isaac. *The Foundation Trilogy: Foundation, Foundation and Empire, Second Foundation*. Gnome Press, 1951.

Ateş, Faruk. "A Brief History of Women in Computing | HackerNoon." August 9, 2017. https://hackernoon.com/a-brief-history-of-women-in-computing-e7253ac24306.

"Attention Span." *GLORIA MARK, PhD*, n.d. Accessed December 3, 2025. https://gloriamark.com/attention-span/.

Averill, James R. Personal Control over Aversive Stimuli and Its Relation to Stress. *Psychological Bulletin* 80, (1973): 286–303.

Bacon, Francis. *The New Organon*. Edited by Lisa Jardin and Michael Silverthorne. Cambridge Texts in the History of Philosophy. Cambridge University Press, 2000.

Bacon, Roger. *Roger Bacon's Letter Concerning the Marvelous Power of Art and of Nature and Concerning the Nullity of Magic*. Chemical Publishing Company, 1923.

Barbour, Sharon, Wright, Nat, and McNamee, Aidan. "NHS Computer Problems Put Patients at Risk of Harm." *BBC News*, May 30, 2024. https://www.bbc.com/news/articles/c4nn0vl2e78o.

Barrett, Claer. "A Dating App with Credit Scores: What's Not to Love? Proving You Can Manage Your Finances Has Real Pulling Power." *Financial Times*, February 10, 2024. https://www.ft.com/content/7ad0c6f5-c3bf-4648-94ad-f3bebccbdf87.

BBC. "BBC World Service – People Fixing The World, Amsterdam's Phone-Free Cafe." August 21, 2024. https://www.bbc.co.uk/programmes/p0jkqfmr.

BBC News. "Busting the Attention Span Myth." Health. March 10, 2017. https://www.bbc.com/news/health-38896790.

BBC News. "Robotic Cheetah 'Breaks Speed Record for Legged Robots.'" Technology. March 6, 2012. https://www.bbc.com/news/technology-17269535.

Beck, Marguerite. "Prosthetic Memory System Successful in Humans, Study Finds." *USC Viterbi | School of Engineering*, March 28, 2025. https://viterbischool.usc.edu/news/2018/03/prosthetic-memory-system-successful-in-humans-study-finds/.

Bednar, Kathrin. "Values and ICTs: A Qualitative Analysis of How New Technologies Affect Us." Master of Science in Cognitive Psychology, WU Vienna, 2018.

Belter, Lisa, Taylor, Jennifer L., Jorgensen, Erica, et al. Newborn Screening and Birth Prevalence for Spinal Muscular Atrophy in the US. *JAMA Pediatrics* 178, no. 9 (2024): 946–949. https://doi.org/10.1001/jamapediatrics.2024.1911.

Bender, Emily M. and Hanna, Alex. *The AI Con: How to Fight Big Tech's Hype and Create the Future We Want*. Random House, 2025.

Berenbach, Brian, and Broy, Manfred. Professional and Ethical Dilemmas in Software Engineering. *IEEE Computer* 1, (2009): 74–80. https://doi.org/10.1109/MC.2009.22.

Berg, Tobias, Burg, Valentin, Gombović, Ana, and Puri, Manju. On the Rise of FinTechs: Credit Scoring Using Digital Footprints. *The Review of Financial Studies* 33, no. 7 (2020): 2845–2897.

Bergson, Henri. *Time and Free Will: An Essay on the Immediate Data of Consciousness*. 1st ed. Routledge, 2014.

Berniker, Mark. "Google Founders' Private Air Terminal Cleared for Takeoff." CNBC, January 16, 2014. https://www.cnbc.com/2014/01/16/google-founders-ok-to-build-private-air-terminal-at-san-jose.html.

Bessière, Katherine, Kiesler, Sara, Kraut, Robert, and Boneva, Bonka S. Effects of Internet Use and Social Resources on Changes in Depression. *Information, Communication & Society* 11, no. 1 (2008): 47–70. https://doi.org/10.1080/13691180701858851.

Big Brother Awards. “Parliamentary Groups of the Christian Democrats (CDU) and the Greens in Hesse.” April 30, 2018. https://bigbrotherawards.de/en/2018/parliamentary-groups-of-the-christian-democrats-cdu-and-the-greens-in-hesse.

Bilton, Nick. “Steve Jobs Was a Low-Tech Parent.” Fashion. *The New York Times*, September 10, 2014. https://www.nytimes.com/2014/09/11/fashion/steve-jobs-apple-was-a-low-tech-parent.html.

Borgmann, Albert. *Technology and the Character of Contemporary Life: A Philosophical Inquiry*. University of Chicago Press, 1984.

Bowker. A ProQuest. Affiliate. “Print ISBN Counts, USA Pubdate 2002-2013.” 2013. https://pq-static-content.proquest.com/collateral/media2/documents/isbn_output_2002_2013.pdf.

Bradley, Sydney, Griffiths, Brent D., and Dixit, Pranav. “Instagram and Facebook are Hardly Social Media Apps Anymore. Here’s the Proof.” Business Insider, April 15, 2025. https://www.businessinsider.com/meta-reveals-decline-friend-content-instagram-facebook-amid-ftc-trial-2025-4.

“Breaking down the Infinite Workday.” June 17, 2025. https://www.microsoft.com/en-us/worklab/work-trend-index/breaking-down-infinite-workday.

Bredow, Gerda. Participatio Singularitatis. Einzigartigkeit als Grundmuster der Weltgestaltung. *Archiv Für Geschichte Der Philosophie* 71, no. 2 (1989): 216–230. https://doi.org/10.1515/agph.1989.71.2.216.

Bregman, Rutger. *Humankind: A Hopeful History*. Bloomsbury Publishing, 2021.

Brehm, Jack W. *A Theory of Psychological Reactance*. Academic Press, 1966.

Brehm, Sharon S. and Brehm, Jack W. *Psychological Reactance: A Theory of Freedom and Control*. Academic Press, 1981.

Brooks, Rodney A. Intelligence without Representation. *Artificial Intelligence* 47, no. 1 (1991): 139–159. https://doi.org/10.1016/0004-3702(91)90053-M.

Bureau of Labor Statistics. “Table 1. Median Years of Tenure with Current Employer for Employed Wage and Salary Workers by Age and Sex, Selected Years, 2014–2024 – 2023 A01 Results.” Accessed November 24, 2025. https://www.bls.gov/news.release/tenure.t01.htm.

CAAAC. “Center for the Alignment of AI Alignment Centers.” Accessed December 4, 2025. https://alignmentalignment.ai.

Campbell, Stefan. “How Many Instagram Followers Does the Average Person Have?” *The Small Business Blog*, October 13, 2024. https://thesmallbusinessblog.com/how-many-instagram-followers-does-the-average-person-have/.

Carr, Nicholas. *The Shallows: What the Internet Is Doing to Our Brains*. W. W. Norton & Company, 2011.

Carraturo, Fabio, Di Perna, Tiziana, Giannicola, Viviana, et al. Envy, Social Comparison, and Depression on Social Networking Sites: A Systematic Review. *European Journal of Investigation in Health, Psychology and Education* 13, no. 2 (2023): 364–376. https://doi.org/10.3390/ejihpe13020027.

Carter, Ian. Positive and Negative Liberty. In *The Stanford Encyclopedia of Philosophy*, Spring 2022, Edited by Edward N. Zalta. Metaphysics Research Lab, Stanford University, 2022. https://plato.stanford.edu/archives/spr2022/entries/liberty-positive-negative/.

Cassirer, Ernst. ‘Philosophy of the Symbolic’ (General Disposition). In *The Genesis of the Symbolic: On the Beginnings of Ernst Cassirer’s Philosophy of Culture*, 1st ed. by Arno Schubbach translated by David J. Hobbs. . De Gruyter, 2021.

Catalanotto, Daniele. “95% of the People Stick to the Default Option.” Medium, February 21, 2019. https://service-design.co/95-of-the-people-stick-to-the-default-option-9e16316a64e1.

O’Grady, Cathleen. “Low-Quality Papers are Surging by Exploiting Public Data Sets and AI.” Science, May 14, 2025. https://www.science.org/content/article/low-quality-papers-are-surging-exploiting-public-data-sets-and-ai.

Chapekis, Athena, and Lieb, Anna. "Google Users are Less Likely to Click on Links When an AI Summary Appears in the Results." *Pew Research Center*, July 22, 2025. https://www.pewresearch.org/short-reads/2025/07/22/google-users-are-less-likely-to-click-on-links-when-an-ai-summary-appears-in-the-results/.

Chatterjee, Rhitu. "Their Teenage Sons Died by Suicide. Now, They are Sounding an Alarm about AI Chatbots." Shots – Health News. *NPR*, September 19, 2025. https://www.npr.org/sections/shots-health-news/2025/09/19/nx-s1-5545749/ai-chatbots-safety-openai-meta-characterai-teens-suicide.

Chayka, Kyle. "Are You Experiencing Posting Ennui?" Infinite Scroll. *The New Yorker*, July 9, 2025. https://www.newyorker.com/culture/infinite-scroll/are-you-experiencing-posting-ennui.

Cheng, Evelyn. "Just 10% of Trading Is Regular Stock Picking, JPMorgan Estimates." Markets. *CNBC*, June 13, 2017. https://www.cnbc.com/2017/06/13/death-of-the-human-investor-just-10-percent-of-trading-is-regular-stock-picking-jpmorgan-estimates.html.

Christensen, Clayton M., Raynor, Michael E., and McDonald, Rory. What Is Disruptive Innovation? *Harvard Business Review*, (2015). https://hbr.org/2015/12/what-is-disruptive-innovation.

Christl, Wolfie, and Spiekermann, Sarah. *Networks of Control: A Report on Corporate Surveillance, Digital Tracking, Big Data & Privacy*. Facultas, 2016.

Clark, Bryan. "Boston Dynamics' New Robot Can Jump 4 Ft, Lift 100 Lbs, and Skate on 2 Wheels." TNW | Insider, February 28, 2017. https://thenextweb.com/news/boston-dynamics-new-robot-can-jump-4-ft-lift-100-lbs-and-skate-on-2-wheels.

Clinton, Virginia. Reading from Paper Compared to Screens: A Systematic Review and Meta-Analysis. *Journal of Research in Reading* 42, no. 2 (2019): 288–325. https://doi.org/10.1111/1467-9817.12269.

Coenen, Christopher. Spotlights on the History of Human Enhancement Discourse. In *The Routledge Handbook of the Ethics of Human Enhancement*. Routledge, 2023.

Coenen, Christopher. Transhumanism in Emerging Technoscience as a Challenge for the Humanities and Technology Assessment. *Teorija in Praksa* 51, no. 5 (2014): 754–771.

Condorcet, Jean-Antoine-Nicolas de Caritat marquis de. The Sketch. In *Condorcet: Political Writings*, Edited by Steven Lukes and Nadia Urbinati. Cambridge University Press, 2012.

Conrad, Sebastian. Enlightenment in Global History: A Historiographical Critique. *The American Historical Review* 117, no. 4 (2012): 999–1027. https://doi.org/10.1093/ahr/117.4.999.

Crawford, Matthew B. *The World beyond Your Head: How to Flourish in an Age of Distraction*. Penguin, 2015.

Csikszentmihalyi, Mihaly. *Flow: The Psychology of Optimal Experience*. Harper Perennial, 1991.

Csikszentmihalyi, Mihaly, and Csikszentmihalyi, Isabella Selega. *Optimal Experience: Psychological Studies of Flow in Consciousness*. University Press, 1988.

Dabbish, Laura, Mark, Gloria, and González, Víctor M. Why Do I Keep Interrupting Myself? Environment, Habit and Self-Interruption. In *Proceedings of the SIGCHI Conference on Human Factors in Computing Systems* (New York, NY, USA), CHI '11, Association for Computing Machinery, May 7, 2011, 3127–3130. https://doi.org/10.1145/1978942.1979405.

Dahlstrom, Daniel O. *Heidegger's Concept of Truth*. Cambridge University Press, 2009.

Damasio, Antonio, and Carvalho, Gil B. The Nature of Feelings: Evolutionary and Neurobiological Origins. *Nature Reviews Neuroscience* 14, no. 2 (2013): 143–152. https://doi.org/10.1038/nrn3403.

Dambra, Savino, Sanchez-Rola, Iskander, Bilge, Leyla, and Balzarotti, Davide. "When Sally Met Trackers: Web Tracking From the Users' Perspective." 2022, 2189–2206. https://www.usenix.org/conference/usenixsecurity22/presentation/dambra.

Delauney, Guy. "World's First AI Minister Will Eliminate Corruption, Says Albania's PM." September 12, 2025. https://www.bbc.com/news/articles/cm2znzgwj3xo.

Delgado, Pablo, Vargas, Cristina, Ackerman, Rakefet, and Salmerón, Ladislao. Don't Throw Away Your Printed Books: A Meta-Analysis on the Effects of Reading Media on Reading Comprehension. *Educational Research Review* 25, (November 2018): 23–38. https://doi.org/10.1016/j.edurev.2018.09.003.

Dembosky, April, and Steel, Emily. "Health Apps Run into Privacy Snags." Financial Times. *Financial Times*, September 1, 2013. https://www.ft.com/content/b709cf4a-12dd-11e3-a05e-00144feabdc0.

Demopoulos, Alaina. "The Women in Love with AI Companions: 'I Vowed to My Chatbot that I Wouldn't Leave Him.'" Technology. *The Guardian*, September 9, 2025. https://www.theguardian.com/technology/2025/sep/09/ai-chatbot-love-relationships.

Descartes, René. *Discourse on Method*. 3rd ed. Translated by Donald A. Cress. Hackett Publishing Company, Inc, 1998.

Devlin, Hannah. "'It's a Real-Life Hunger Games': A Lifesaving Drug Costs $2m, but Not Every Child Can Get It." Business. *The Guardian*, March 5, 2022. https://www.theguardian.com/business/2022/mar/05/real-life-hunger-games-lifesaving-drug-costs-2m-dollars.

"DGB-Mitgliederzahlen." Accessed September 23, 2025. https://www.dgb.de/der-dgb/geschichte-des-dgb/mitgliederzahlen/.

Döpfner, Mathias. "Warum Wir Google Fürchten." FAZ, April 16, 2014. https://www.faz.net/aktuell/feuilleton/medien-und-film/mathias-doepfner-warum-wir-google-fuerchten-12897463.html.

"Drones Have Become the Dominant Killers in the Ukraine War – National Security News". April 9, 2025. https://nationalsecuritynews.com/2025/04/drones-have-become-the-dominant-killers-in-the-ukraine-war/.

Dunning, David. The Dunning–Kruger Effect: On Being Ignorant of One's Own Ignorance. In *Advances in Experimental Social Psychology*, Vol. 44. Academic Press, 2011. https://doi.org/10.1016/B978-0-12-385522-0.00005-6.

Edwards, Chris. "Sask. Mushroom Pickers Sound Alarm on Dangerously Misleading AI-Generated Books." *CBC News*, April 5, 2025. https://www.cbc.ca/news/canada/saskatchewan/mushroom-picking-ai-generated-1.7502667.

Edwards, Steven M., Li, Hairong, and Lee, Joo-Hyun. Forced Exposure and Psychological Reactance: Antecedents and Consequences of the Perceived Intrusiveness of Pop-Up Ads. *Journal of Advertising* 31, no. 3 (2002): 83–95. https://doi.org/10.1080/00913367.2002.10673678.

Eisemann, Nora, Bunk, Stefan, Mukama, Trasias, et al. Nationwide Real-World Implementation of AI for Cancer Detection in Population-Based Mammography Screening. *Nature Medicine* 31, no. 3 (2025): 917–924. https://doi.org/10.1038/s41591-024-03408-6.

Engel, Christoph, Linhardt, Lorenz, and Schubert, Marcel. Code Is Law: How COMPAS Affects the Way the Judiciary Handles the Risk of Recidivism. *Artificial Intelligence and Law* 33, no. 2 (2025): 383–404. https://doi.org/10.1007/s10506-024-09389-8.

Engels, Friedrich *Anti-Dühring. Herr Eugen Dühring's Revolution in Science*. Progress Publishers, 1947. http://archive.org/details/antiduhring.

Etchells, Pete, and Chambers, Chris. "Mindless Eating: Is There Something Rotten behind the Research? | Science | The Guardian." *The Guardian*, February 16, 2018. https://www.theguardian.com/science/head-quarters/2018/feb/16/mindless-eating-brian-wansink-is-there-something-rotten-behind-the-research.

Eurostat. "[Ilc_lvph02] Distribution of Households by Household Type from 2003 Onwards." 2025. https://ec.europa.eu/eurostat/databrowser/view/ilc_lvph02__custom_16784682/bookmark/table?bookmarkId=f3ebff55-af69-4df8-8e56-ba595c34824c&c=1747742464563&page=time:2024.

"EXPLANATORY NOTES Post Office (Horizon System) Offences Act 2024 Chapter 14." n.d. Accessed December 3, 2025. https://www.legislation.gov.uk/ukpga/2024/14/pdfs/ukpgaen_20240014_en.pdf.

Exploding Topics. "Time Spent Using Smartphones (2025 Statistics)." March 28, 2025. https://explodingtopics.com/blog/smartphone-usage-stats.

Exploding Topics. "Worldwide Daily Social Media Usage (New 2025 Data)." May 4, 2022. https://explodingtopics.com/blog/social-media-usage.

Eyal, Nir. *Hooked: How to Build Habit-Forming Products*. Penguin Canada, 2014.

"Facebook and Instagram to Get £2.99 UK Subscription to Stop Ads." September 26, 2025. https://www.bbc.com/news/articles/c5y4xl5x8q7o.

"Fairwork Cloudwork Ratings 2023: Work in the Planetary Labour Market." Accessed September 10, 2025. https://fair.work/en/fw/publications/fairwork-cloudwork-ratings-2023-work-in-the-planetary-labour-market/.

FBTB, prod. *Alexander Osterwalder – Value Proposition Design*. 2017. 58:51. https://www.youtube.com/watch?v=b_X18bmpHaw.

Forster, Georg. Über lokale und allgemeine Bildung. In *Kleine Schriften zu Kunst und Literatur: Sakontala*, Edited by Gerhard Steiner. Georg Forsters Werke: sämtliche Schriften, Tagebücher, Briefe. 7. De Gruyter, 1990.

Foucault, Michel. *History of Madness*. 1st ed. Translated by Jean Khalfa and Jonathan Murphy. Routledge, 2013.

Foucault, Michel. On the Genealogy of Ethics: An Overview of Work in Progress. In *The Essential Works of Foucault, 1954-1984: Aesthetics, Method, and Epistemology*. New Press, 1997.

Frenkel, Edward. "Google Should Not Be Allowed to Secretly Collect Private Medical Data." Tech. *HuffPost*, May 20, 2016. https://www.huffpost.com/entry/google-should-not-be-allo_b_10057602.

Frey, Carl Benedikt and Osborne, Michael A. The Future of Employment: How Susceptible are Jobs to Computerisation? *Technological Forecasting and Social Change* 114, ((January 2017): 254–280. https://doi.org/10.1016/j.techfore.2016.08.019.

Frost, Jeana H., Chance, Zoë, Norton, Michael I., and Ariely, Dan. People are Experience Goods: Improving Online Dating with Virtual Dates. *Journal of Interactive Marketing* 22, no. 1 (2008): 51–61. https://doi.org/10.1002/dir.20107.

Führer, Markus. Albert the Great. In *The Stanford Encyclopedia of Philosophy*, Summer 2022, Edited by Edward N. Zalta. Metaphysics Research Lab, Stanford University, 2022. https://plato.stanford.edu/archives/sum2022/entries/albert-great/.

Fürstenau, Marcel. "German Police Expands Use of Palantir Surveillance Software." Dw.Com, April 8, 2025. https://www.dw.com/en/german-police-expands-use-of-palantir-surveillance-software/a-73497117.

Future of Humanity Institute. "Future of Humanity Institute." Accessed December 4, 2025. https://www.futureofhumanityinstitute.org.

Gabrielle, Vincent. "How Employers Have Gamified Work for Maximum Profit | Aeon Essays." *Aeon*, October 10, 2018. https://aeon.co/essays/how-employers-have-gamified-work-for-maximum-profit.

Gartner. "Gartner 2024 Hype Cycle for Emerging Technologies Highlights Developer Productivity, Total Experience, AI and Security." August 21, 2024. https://www.gartner.com/en/newsroom/press-releases/2024-08-21-gartner-2024-hype-cycle-for-emerging-technologies-highlights-developer-productivity-total-experience-ai-and-security.

Gebru, Timnit and Torres, Émile P. "The TESCREAL Bundle: Eugenics and the Promise of Utopia through Artificial General Intelligence." *First Monday*, ahead of print, April 14, 2024. https://doi.org/10.5210/fm.v29i4.13636.

Gehring, Petra. Wissen. In *Historisches Wörterbuch der Philosophie*, Vol. 12, Edited by Joachim Ritter. Schwabe & Co., 2004.

Ghosh, Shona. "Undercover Author Finds Amazon Warehouse Workers in UK 'Peed in Bottles' over Fears of Being Punished for Taking a Break." *Business Insider*. Accessed September 10, 2025. https://www.businessinsider.com/amazon-warehouse-workers-have-to-pee-into-bottles-2018-4.

Gibson, James Jerome. *The Ecological Approach to Visual Perception*. Houghton Mifflin, 1979.

Glanvill, Joseph G. *Plus Ultra or the Progress and Advancement of Knowledge since the Days of Aristotle*. James Collins, 1668.

Glenberg, Arthur M. What Memory Is For. *Behavioral and Brain Sciences* 20, no. 1 (1997): 1–19. https://doi.org/10.1017/S0140525X97000010.

Goddard, Emily, Marlon, Jennifer, Ballew, Matthew, et al. "Explore 16 Years of U.S. Climate Opinions with the CCAM Explorer." *Yale Program on Climate Change Communication* (New Haven, CT), May 6, 2025. https://climatecommunication.yale.edu/publications/explore-16-years-of-u-s-climate-opinions-with-the-ccam-explore/.

Governing AI for Humanity. United Nations, 2024. https://www.un.org/sites/un2.un.org/files/governing_ai_for_humanity_final_report_en.pdf.

Guangheng, Wu. *The Subscription Prices for 55% Elsevier Journals Increase*. 2024. https://doi.org/10.13140/RG.2.2.19347.69926.

Guenther, Oliver, and Spiekermann, Sarah. RFID and Perceived Control: The Consumer's View. *Communications of the ACM* 48, no. 9 (2005): 73–76.

Haddon, Heather. "Rise of the Restaurant Robots: Chipotle, Sweetgreen and Others Bet on Automation." *The Wall Street Journal*, July 11, 2024. https://www.wsj.com/business/hospitality/restaurant-robots-flippy-chippy-autocado-9de44eeb.

Halmer, Nikolaus. "Michel Maffesoli: Orgiastische Wissenschaft – Science.ORF.At." November 14, 2014. https://sciencev2.orf.at/stories/1749443/index.html.

Halpern, Sue. Review of *Mind Control & the Internet*, by Michael Chorost, Eli Pariser, and Jaron Lanier. *The New York Review of Books*, June 23, 2011. https://www.nybooks.com/articles/2011/06/23/mind-control-and-internet/.

Hamann, Florian. "Karriereturbo oder Ruin: Was der Doktortitel für den beruflichen Erfolg in Banking und Consulting bringt." eFinancialCareers, March 13, 2019. https://www.efinancialcareers.de/nachrichten/2018/03/karriereturbo-oder-ruin-was-der-doktortitel-fur-den-beruflichen-erfolg-in-banking-und-consulting-bringt.

Hao, Karen. *Empire of AI: Inside the Reckless Race for Total Domination*. Allen Lane, 2025.

Hara, Kotaro, Adams, Abigail, Milland, Kristy, Savage, Saiph, Callison-Burch, Chris, and Bigham, Jeffrey P. A Data-Driven Analysis of Workers' Earnings on Amazon Mechanical Turk. In *Proceedings of the 2018 CHI Conference on Human Factors in Computing Systems* (New York, NY, USA), CHI '18, Association for Computing Machinery. April 21, 2018, 1–14. https://doi.org/10.1145/3173574.3174023.

Haraway, Donna J. A Cyborg Manifesto: Science, Technology, and Socialist-Feminism in the 1980s. *Socialist Review* 15, (1985).

Harding, William. *AI Copilot Code Quality*. GitClear, 2025. https://gitclear-public.s3.us-west-2.amazonaws.com/GitClear-AI-Copilot-Code-Quality-2025.pdf.

Harrison, Robert Pogue. *Juvenescence: A Cultural History of Our Age*. University of Chicago Press, 2022.

Helft, Miguel. “A New (Fighter) Jet for Google’s Founders?” Bits Blog, October 23, 2008. https://archive.nytimes.com/bits.blogs.nytimes.com/2008/10/23/a-new-fighter-jet-for-googles-founders/.

Helm, Toby. “Revealed: How Drugs Giants Can Access Your Health Records.” Technology. *The Guardian*, February 8, 2020. https://www.theguardian.com/technology/2020/feb/08/fears-over-sale-anonymous-nhs-patient-data.

Henton, Doug, and Held, Kim. The Dynamics of Silicon Valley: Creative Destruction and the Evolution of the Innovation Habitat. *Social Science Information* 52, no. 4 (2013): 539–557. https://doi.org/10.1177/0539018413497542.

Heritage, Stuart. “‘I Felt Pure, Unconditional Love’: The People Who Marry Their AI Chatbots.” Television & Radio. *The Guardian*, July 12, 2025. https://www.theguardian.com/tv-and-radio/2025/jul/12/i-felt-pure-unconditional-love-the-people-who-marry-their-ai-chatbots.

Hern, Alex. “Sci-Fi Publisher Clarkesworld Halts Pitches amid Deluge of AI-Generated Stories.” Technology. *The Guardian*, February 21, 2023. https://www.theguardian.com/technology/2023/feb/21/sci-fi-publisher-clarkesworld-halts-pitches-amid-deluge-of-ai-generated-stories.

Herrera, Sebastian. “Exclusive | Amazon Is on the Cusp of Using More Robots than Humans in Its Warehouses.” *The Wall Street Journal*. Accessed September 10, 2025. https://www.wsj.com/tech/amazon-warehouse-robots-automation-942b814f.

Heyman, Tom, Moors, Pieter, and Storms, Gert. On the Cost of Knowledge: Evaluating the Boycott against Elsevier. *Frontiers in Research Metrics and Analytics* 1, (October 2016). https://doi.org/10.3389/frma.2016.00007.

Hobbes, Thomas. *Leviathan*. Hackett, 1994.

Hobsbawm, Julia. *Fully Connected: Social Health in an Age of Overload*. Bloomsbury Business, 2018.

Hodson, Hal. “Revealed: Google AI Has Access to Huge Haul of NHS Patient Data.” *NewScientist*, April 29, 2016. https://www.newscientist.com/article/2086454-revealed-google-ai-has-access-to-huge-haul-of-nhs-patient-data/.

Hoff, Johannes. Iconicity and the Anamorphosis of Social Space: Retrieving Nicholas of Cusa’s Political Pneumatology. In *Veni, Sancte Spiritus! Theologische Beiträge Zur Sendung Des Geistes. Contributions to the Mission of the Spirit*, Edited by Walter Dürr and Guido Vergauwen. Aschendorff, 2018.

Hoff, Johannes. *The Analogical Turn: Rethinking Modernity with Nicholas of Cusa*. William B. Erdmans Publishing Company, 2013.

Holmes, Catherine. *Report: Royal Commission into the Robodebt Scheme*. Royal Commission into the Robodebt Scheme, 2023.

Hoofnagle, Chris Jay, Urban, Jennifer M., and Li, Su. “Privacy and Modern Advertising: Most US Internet Users Want ‘Do Not Track’ to Stop Collection of Data about Their Online Activities.” SSRN Scholarly Paper No. 2152135. Social Science Research Network, October 8, 2012. https://papers.ssrn.com/abstract=2152135.

Horvitz, Eric J., Hovel, David O., Jacobs, Andrew W., and Kadie, Carl M. Notification Platform Architecture. World Intellectual Property Organization Patent WO2001069387A8, filed March 16, 2001, and issued December 6, 2001. https://patents.google.com/patent/WO2001069387A8/en.

Horvitz, Eric, Kadie, Carl, Paek, Tim, and Hovel, David. Models of Attention in Computing and Communication: From Principles to Applications. *Communications of the ACM* 46, no. 3 (2003): 52–59. https://doi.org/10.1145/636772.636798.

Huang, Saffron, Durmus, Esin, McCain, Miles, et al. Values in the Wild: Discovering and Analyzing Values in Real-World Language Model Interactions. *arXiv:2504.15236. Preprint, arXiv*, (April 21, 2025). https://doi.org/10.48550/arXiv.2504.15236.

Humanity+. "Transhumanist FAQ." Accessed December 4, 2025. https://www.humanityplus.org/transhumanist-faq.

Hume, David. *Enquiries Concerning Human Understanding and Concerning the Principles of Morals.* 3d ed. Edited by Lewis Amherst Selby-Bigge and Peter Harold Nidditch. Clarendon Press, 1975.

iBlindGame, prod. *Inattentional Blindness – The Monkey Business Illusion*. 2011. 1:41. https://www.youtube.com/watch?v=f94o3B3csYI.

"IEEE – The World's Largest Technical Professional Organization Dedicated to Advancing Technology for the Benefit of Humanity. | IEEE." Accessed December 4, 2025. https://www.ieee.org/.

"International Publishing Data 2023." *International Publishers Association*, November 12, 2023. https://internationalpublishers.org/international-publishing-data-2023/.

Irani, Lilly. Difference and Dependence among Digital Workers: The Case of Amazon Mechanical Turk. *South Atlantic Quarterly* 114, no. 1 (2015): 225–234. https://doi.org/10.1215/00382876-2831665.

Isaacson, Walter. *Steve Jobs*. Simon & Schuster, 2011.

Jaeger, Johannes, Riedl, Anna, Djedovic, Alex, Vervaeke, John, and Walsh, Denis. Naturalizing Relevance Realization: Why Agency and Cognition are Fundamentally Not Computational. *Frontiers in Psychology* 15, (June 2024). https://doi.org/10.3389/fpsyg.2024.1362658.

Jagtap, Suraj. "Mobile Application Statistics and Facts." *Market.Biz*, September 25, 2025. https://market.biz/mobile-application-statistics/.

Jobin, Anna, Ienca, Marcello, and Vayena, Effy. The Global Landscape of AI Ethics Guidelines. *Nature Machine Intelligence* 1, no. 9 (2019): 389–399. https://doi.org/10.1038/s42256-019-0088-2.

Joyce, James. *Ulysses*. Edited by Jeri Johnson. Oxford University Press, 1998.

JuchemGruppe, prod. *Qkies Sag´s Mit Keksen*. 2011. 2:17. https://www.youtube.com/watch?v=r06_vJUWHJg.

Kahneman, Daniel. *Attention and Effort*. Prentice-Hall, 1973.

Kant, Immanuel. *Groundwork of the Metaphysics of Morals*. Edited by Mary Gregor and Jens Timmermann. Cambridge University Press, 2012.

Kaye, Jane, Whitley, Edgar A., Lund, David, Morrison, Michael, Teare, Harriet, and Melham, Karen. Dynamic Consent: A Patient Interface for Twenty-First Century Research Networks. *European Journal of Human Genetics* 23, no. 2 (2014): 141–146. https://doi.org/10.1038/ejhg.2014.71.

Kelly, Eugene. *Material Ethics of Value: Max Scheler and Nicolai Hartmann*. Vol. 203. Phaenomenologica. Springer Netherlands, 2011. https://doi.org/10.1007/978-94-007-1845-6.

Kerscher, Klaus-Jürgen. *Homo oeconomicus und Menschenbild: Form und Wesen einer beachtenswerten Spannung*. Metropolis-Verl, 2013.

Khamassi, Mehdi, Nahon, Marceau, and Chatila, Raja. Strong and Weak Alignment of Large Language Models with Human Values. *Scientific Reports* 14, no. 1 (2024): 19399. https://doi.org/10.1038/s41598-024-70031-3.

Knight, Will. The Dark Secret at the Heart of AI. *MIT Technology Review* 120, no. 3 (2017): 54–63.

Kohavi, Ronny, Crook, Thomas, Longbotham, Roger, et al. "Online Experimentation at Microsoft." July 2009. https://www.microsoft.com/en-us/research/publication/online-experimentation-at-microsoft/.

Kolhatkar, Sheelah. "Welcoming Our New Robot Overlords." Annals of Technology. *The New Yorker*, October 16, 2017. https://www.newyorker.com/magazine/2017/10/23/welcoming-our-new-robot-overlords.

Kosmyna, Nataliya, Hauptmann, Eugene, Yuan, Ye Tong, et al. Your Brain on ChatGPT: Accumulation of Cognitive Debt When Using an AI Assistant for Essay Writing Task. *Preprint, Arxiv*, (June 10, 2025). https://doi.org/10.48550/arXiv.2506.08872.

Krasnova, Hanna, Widjaja, Thomas, Buxmann, Peter, Wenninger, Helena, and Benbasat, Izak. Why Following Friends Can Hurt You: An Exploratory Investigation of the Effects of Envy on Social Networking Sites among College-Age Users (US) . *Information Systems Research* 26, no. 3 (2015): 585–605. https://doi.org/10.1287/isre.2015.0588.

Krempl, Stefan. "Palantir: German Investigators Also Use Big Data for Minor Crimes." *Heise Online*, June 19, 2025. https://www.heise.de/en/news/Palantir-German-investigators-also-use-big-data-for-minor-crimes-10453589.html.

Krysmanski, Hans Jürgen. Der Ganz Alltägliche Transhumanismus. *Jahrbuch Für Pädagogik* 2014, no. 1 (2014): 123–142. https://doi.org/10.3726/265764_123.

Kurzweil, Ray. *The Singularity Is Near: When Humans Transcend Biology*. Penguin, 2005.

Kushlev, Kostadin, and Dunn, Elizabeth W. Checking Email Less Frequently Reduces Stress. *Computers in Human Behavior* 43, (February 2015): 220–228. https://doi.org/10.1016/j.chb.2014.11.005.

Langer, Ellen. *The Psychology of Control*. Sage Publications, 1983.

Lawrence, Felicity. "Uber Is Treating Its Drivers as Sweated Labour, Says Report." Technology. *The Guardian*, December 9, 2016. https://www.theguardian.com/technology/2016/dec/09/uber-drivers-report-sweated-labour-minimum-wage.

Lee, Hao-Ping (Hank), Sarkar, Advait, Tankelevitch, Lev, et al. "The Impact of Generative AI on Critical Thinking: Self-Reported Reductions in Cognitive Effort and Confidence Effects from a Survey of Knowledge Workers." Paper presented at CHI 2025. April 1, 2025. https://www.microsoft.com/en-us/research/publication/the-impact-of-generative-ai-on-critical-thinking-self-reported-reductions-in-cognitive-effort-and-confidence-effects-from-a-survey-of-knowledge-workers/.

Lemmens, Pieter. 'This System Does Not Produce Pleasure Anymore'. An Interview with Bernard Stiegler. *Krisis | Journal for Contemporary Philosophy* 31, no. 1 (2011): 33–41.

Klapper, Leora, Singer, Dorothe, Starita, Laura, and Norris, Alexandra. *The Global Findex Database 2025*. World Bank Group, 2025. https://www.worldbank.org/en/publication/globalfindex.

Leroi-Gourhan, André. *Gesture and Speech*. MIT Press, 1993.

Levin, Benjamin. A Defensible Defense?: Reexamining Castle Doctrine Statutes. *Harvard Journal on Legislation*, (January 1, 2010). https://scholar.law.colorado.edu/faculty-articles/450.

Lieberman, Matthew D. *Social: Why Our Brains are Wired to Connect*. Oxford University Press, 2013.

MacIntyre, Alasdair. *After Virtue: A Study in Moral Theory*. University of Notre Dame Press, 1984.

Making Change Work. IBM, 2008. https://public.dhe.ibm.com/software/be/Making_Change_Work_eff.pdf.

Maslow, Abraham H. *Motivation and Personality*. 2d ed. Harper & Row, 1970.

Mazzucato, Mariana. *The Value of Everything: Making and Taking in the Global Economy*. Allen Lane, 2018.

McClain, Colleen and Gelles-Watnick, Risa. "From Looking for Love to Swiping the Field: Online Dating in the U.S." *Pew Research Center*, February 2, 2023. https://www.pewresearch.org/internet/2023/02/02/from-looking-for-love-to-swiping-the-field-online-dating-in-the-u-s/.

McClelland, David C. *Human Motivation*. Cambridge University Press, 1988. https://doi.org/10.1017/CBO9781139878289.

McCright, Aaron M. and Dunlap, Riley E. The Politicization of Climate Change and Polarization in the American Public's Views of Global Warming, 2001–2010. *The Sociological Quarterly* 52, no. 2 (2011): 155–194. https://doi.org/10.1111/j.1533-8525.2011.01198.x.

McGilchrist, Iain. *The Master and His Emissary: The Divided Brain and the Making of the Western World*. 1st ed. Yale University Press, 2009. https://doi.org/10.12987/9780300247459.

McHugh, Molly. “Facebook Can Recognize You Even If You Don’t Show Your Face.” Tags. *Wired*, n.d. Accessed December 4, 2025. https://www.wired.com/2015/06/facebook-can-recognize-even-dont-show-face/.

Metzger, Chloe. “You’re so Vain: Inside the Obsession with Our Own Social Media Profiles.” *Marie Claire*, February 12, 2018. https://www.marieclaire.com/culture/a16767138/social-media-narcissism/.

Milbank, John, and Pabst, Adrian. *The Politics of Virtue: Postliberalism and the Human Future*. Rowman & Littlefield, 2016.

Montgomery, Blake. “Mother Says AI Chatbot Led Her Son to Kill Himself in Lawsuit against Its Maker.” Technology. *The Guardian*, October 23, 2024. https://www.theguardian.com/technology/2024/oct/23/character-ai-chatbot-sewell-setzer-death.

More, Max and Vita-More, Natasha. *The Transhumanist Reader: Classical and Contemporary Essays on the Science, Technology, and Philosophy of the Human Future*. John Wiley & Sons, 2013.

Morozov, Evgeny. *To Save Everything, Click Here*. Penguin, 2013.

Morris, Cora. “Death on Delivery: New Report Examines Civilian Harm from Intensifying Drone Warfare across Africa.” *Drone Wars UK*, March 10, 2025. https://dronewars.net/2025/03/10/death-on-delivery-new-report-examines-civilian-harm-from-intensifying-drone-warfare-across-africa/.

NBC News. “Is Your TV Watching You? Latest Models Raise Concerns.” March 19, 2012. http://www.nbcnews.com/technolog/your-tv-watching-you-latest-models-raise-concerns-483619.

“New Texas Law Bans Student Personal Device Use during School Hours (En Español) | Austin ISD.” August 13, 2025. https://www.austinisd.org/announcements/2025/08/13/new-texas-law-bans-student-personal-device-use-during-school-hours-en.

Nguyen, C Thi. “Escape the Echo Chamber.” *Aeon*, April 9, 2018. https://aeon.co/essays/why-its-as-hard-to-escape-an-echo-chamber-as-it-is-to-flee-a-cult.

Nhs.Uk. “Spinal Muscular Atrophy (SMA).” August 30, 2024. https://www.nhs.uk/conditions/spinal-muscular-atrophy-sma/.

Ni, Liangkang *Zur Sache des Bewusstseins: Phänomenologie – Buddhismus – Konfuzianismus*. Königshausen u. Neumann, 2010.

Noble, D. F. *The Religion of Technology, The Divinity of Man and the Spirit of Invention*. Penguin, 1999.

Nonaka, Ikujiro, and Takeuchi, Hirotaka. The Wise Leader. *Harvard Business Review*, (2011): 58–67.

Nonaka, Ikujirō, Toyama, Ryoko, and Hirata, Toru. eds. *Managing Flow: A Process Theory of the Knowledge-Based Firm*. Palgrave Macmillan, 2008.

Norton, Michael I. Less Is More: The Lure of Ambiguity or Why Familiarity Breeds Contempt. *Journal of Personality and Social Psychology* 92, no. 1 (2007): 97–105.

OECD. “Science, Technology and Innovation Scoreboard.” Accessed December 4, 2025. https://www.oecd.org/en/data/datasets/science-technology-and-innovation-scoreboard.html.

Ofcom. “Children’s Media Lives.” April 19, 2023. https://www.ofcom.org.uk/media-use-and-attitudes/media-habits-children/childrens-media-lives.

OpenAI Help Center. “GPT-5.1 in ChatGPT.” Accessed December 4, 2025. https://help.openai.com/en/articles/11909943-gpt-51-in-chatgpt#h_1fadb43e65.

Pariser, Eli. *The Filter Bubble: What The Internet Is Hiding From You*. Penguin UK, 2011.

Park, Michael, Leahey, Erin, and Funk, Russell J. Papers and Patents are Becoming Less Disruptive over Time. *Nature* 613, no. 7942 (2023): 138–144. https://doi.org/10.1038/s41586-022-05543-x.

Pascoe, Robin. “Over 33,000 Families Acknowledged as Benefit Scandal Victims.” *DutchNews.Nl*, January 22, 2024. https://www.dutchnews.nl/2024/01/over-33000-families-acknowledged-as-benefit-scandal-victims/.

Plato. *Texte zur Ideenlehre*. Edited and translated by Hans-Georg Gadamer. V. Klostermann, 1978.

Polanyi, Michael. *The Tacit Dimension*. Anchor Books, 1967.

Porter, Michael E. *Competitive Strategy: Techniques for Analyzing Industries and Competitors*. Free Press, 2008.

Porter, Michael E., and Kramer, Mark R. "Creating Shared Value." Corporate Social Responsibility. *Harvard Business Review*, January 1, 2011. https://hbr.org/2011/01/the-big-idea-creating-shared-value.

Quinio, Akila. "What are Banks Doing with Your Financial Data?" Personal Finance. *Financial Times*, November 22, 2024. https://www.ft.com/content/754e598e-e96d-4a32-b4d2-a949f762b537.

Redman, Thomas C. Data's Credibility Problem. *Harvard Business Review*, (2013).

Regalado, Antonio. "Researchers are Keeping Pig Brains Alive Outside the Body." MIT Technology Review, April 25, 2018. https://www.technologyreview.com/2018/04/25/240742/researchers-are-keeping-pig-brains-alive-outside-the-body/.

Reinsberg, Alexandra, and Weiske, Cosima. *Interview mit einem Foodora-Lieferanten: "Es ist ein Knochenjob."* November 8, 2016. https://www.br.de/puls/themen/leben/arbeiten-fuer-foodora-so-ist-es-wirklich-interview-100.html.

Reiss, Steven. Multifaceted Nature of Intrinsic Motivation: The Theory of 16 Basic Desires. *Review of General Psychology* 8, no. 3 (2004): 179–193.

Renstrom, Joelle. "What Happened When I Made My Students Turn off Their Phones | Aeon Ideas." *Aeon*, October 18, 2017. https://aeon.co/ideas/what-happened-when-i-made-my-students-turn-off-their-phones.

Ritter, Joachim. *Historisches Worterbuch der Philosophie*. Vol. 2. Schwabe & Co. Verlag, 1972.

Rousseau, Jean-Jacques. *Discourse on the Origin of Inequality*. Translated by Donald A. Cress. Hackett Publishing Company, Inc, 1992.

Runge, Tobias, Thüm, Thomas, Cleophas, Loek, Schaefer, Ina, and Watson, Bruce W. Comparing Correctness-by-Construction with Post-Hoc Verification – A Qualitative User Study. In *Formal Methods. FM 2019 International Workshops*, Edited by Emil Sekerinski, Nelma Moreira, José N. Oliveira, et al. Springer International Publishing, 2020. https://doi.org/10.1007/978-3-030-54997-8_25.

Sanou, Hanneke. "Over 1,100 Children of Childcare Benefit Victims Taken into Care." *DutchNews.Nl*, October 20, 2021. https://www.dutchnews.nl/2021/10/over-1100-children-of-childcare-benefit-victims-taken-into-care/.

Scheler, Max. *Formalism in Ethics and Non-Formal Ethics of Values: A New Attempt Toward the Foundation of an Ethical Personalism*. Translated by Manfred S. Frings and Roger L. Funk. Northwestern University Press, 1973. http://archive.org/details/formalisminethic0000sche.

Seiser, Birgit. "350 Ärzte verkaufen Daten von Patienten." *Kurier*, August 19, 2013. https://kurier.at/chronik/oesterreich/350-aerzte-verkaufen-daten-von-patienten/23.324.371.

Semuels, Alana. "The Internet Is Enabling a New Kind of Poorly Paid Hell." Business. *The Atlantic*, January 23, 2018. https://www.theatlantic.com/business/archive/2018/01/amazon-mechanical-turk/551192/.

Sen, Amartya. Capability and Well-Being. In *The Quality of Life*, Edited by Martha Nussbaum and Amartya Sen. Oxford University Press, 1993. https://doi.org/10.1093/0198287976.003.0003.

Sexual Extortion & Young People: Navigating Threats in Digital Environments. Thorn, 2025. https://info.thorn.org/hubfs/Research/Thorn_SexualExtortionandYoungPeople_June2025.pdf.

Shapiro, Lawrence. *Embodied Cognition*. Routledge, 2011.

Sherman, Mark. Using ChatGPT to Analyze Your Code? Not so Fast. February 12, (2024). https://insights.sei.cmu.edu/blog/using-chatgpt-to-analyze-your-code-not-so-fast/.

Shotter, James. "A Taste for Risk and Something Different." *Financial Times*, November 20, 2012. https://www.ft.com/content/31a2f6cc-3243-11e2-b891-00144feabdc0.

Simon, Felix, Nielsen, Rasmus Kleis, and Fletcher, Richard. *Generative AI and News Report 2025: How People Think about AI's Role in Journalism and Society*. Reuters Institute, University of Oxford, 2025. doi:10.60625/risj-5bjv-yt69.

Simonite, Tom. "Moore's Law Is Dead. Now What?" *MIT Technology Review*, May 13, 2016. https://www.technologyreview.com/2016/05/13/245938/moores-law-is-dead-now-what/.

Simons, Daniel J. Inattentional Blindness. *Scholarpedia* 2, no. 5 (2007): 3244. https://doi.org/10.4249/scholarpedia.3244.

Singer, P. W., and Brooking, Emerson T. *Likewar: The Weaponization of Social Media*. Houghton Mifflin Harcourt, 2018.

Singh, Namita. "Woman 'Weds' AI Persona She Created on ChatGPT." Asia. *The Independent*, November 13, 2025. https://www.independent.co.uk/asia/japan/japan-chatgpt-ai-wedding-b2864314.html.

Singleton, Tom, Gerken, Tom, and McMahon, Liv. "How a Chatbot Encouraged a Man Who Wanted to Kill the Queen." *BBC News*, October 6, 2023. https://www.bbc.com/news/technology-67012224.

"Singularity | Leading Innovation & Exponential Technology Education." Accessed December 4, 2025. https://www.su.org.

"Sir Edward Coke Declares That Your House Is Your 'Castle and Fortress'." Accessed December 4, 2025. https://oll.libertyfund.org/quotes/502.

Smith, Benedict, and Kelly, Kieran. "A 14-Year-Old Boy Fell in Love with a Flirty AI Chatbot. He Shot Himself so They Could Die Together." *The Telegraph*, October 24, 2024. https://www.telegraph.co.uk/us/news/2024/10/24/teenage-boy-killed-himself-fall-love-ai-chatbot/.

Solon, Olivia. "Deus Ex Machina: Former Google Engineer Is Developing an AI God." Technology. *The Guardian*, September 28, 2017. https://www.theguardian.com/technology/2017/sep/28/artificial-intelligence-god-anthony-levandowski.

Solove, Daniel J. A Taxonomy of Privacy. *University of Pennsylvania Law Review* 154, no. 3 (2006): 477–560.

Somers, James. "The Coming Software Apocalypse." Technology. *The Atlantic*, September 26, 2017. https://www.theatlantic.com/technology/archive/2017/09/saving-the-world-from-code/540393/.

"SONNENTOR Organic Teas & Organic Spices." Accessed September 10, 2025. https://www.sonnentor.com/en-gb/about-us/history.

Spiekermann, Sarah, and Pallas, Florian. Technology Paternalism – Wider Implications of Ubiquitous Computing. *Poiesis & Praxis* 4, no. 1 (2006): 6–18. https://doi.org/10.1007/s10202-005-0010-3.

Spiekermann, Sarah. *Ethical IT Innovation: A Value-Based System Design Approach*. Taylor & Francis, 2016.

Spiekermann, Sarah, Hampson, Peter, Ess, Charles M., Hoff, Johannes, Coeckelbergh, Mark, and Franck, Georg. "The Ghost of Transhumanism & the Sentience of Existence." SSRN Scholarly Paper No. 3617635. Social Science Research Network, June 19, 2017. https://papers.ssrn.com/abstract=3617635.

Spiekermann, Sarah, and Korunovska, Jana. The Importance of Interface Complexity and Entropy for Online Information Sharing. *Behaviour & Information Technology* 33, no. 6 (2014): 636–645. https://doi.org/10.1080/0144929X.2013.845910.

Spiekermann, Sarah, Korunovska, Jana, and Langheinrich, Marc. Inside the Organization: Why Privacy and Security Engineering Is a Challenge for Engineers. *Proceedings of the IEEE* 107, no. 3 (2019): 600–615. https://doi.org/10.1109/JPROC.2018.2866769.

STADA. “STADA Health Report 2025.” 2025. https://www.stada.com/media/health-reports/stada-health-report-2025.

State of the Global Workplace Report. Gallup, 2025. https://www.gallup.com/workplace/349484/state-of-the-global-workplace.aspx.

Statista. “Ausbildungsberufe in Deutschland bis 2023.” Accessed December 4, 2025. https://de.statista.com/statistik/daten/studie/156901/umfrage/ausbildungsberufe-in-deutschland/.

Statista. “Number of Apprentices in Germany from 1950 to 2023.” Accessed December 4, 2025. https://www.statista.com/statistics/1182007/number-vocational-trainees-germany/.

Statista. “Österreich – Beste Gewürzmarken 2024.” Accessed December 3, 2025. https://de.statista.com/statistik/daten/studie/1405676/umfrage/ranking-der-besten-marken-fuer-gewuerze-in-oesterreich/.

Statista. “Population Numbers in Germany 2024.” Accessed September 23, 2025. https://www.statista.com/statistics/672608/development-population-numbers-germany/.

Stiegler, Bernard. *States of Shock: Stupidity and Knowledge in the 21st Century*. Translated by Daniel Ross. Polity, 2015.

Strickland, Eliza. “Treating Depression with tDCS: Startup Ybrain Aims for the Mainstream.” *IEEE Spectrum*, January 26, 2017. https://spectrum.ieee.org/treating-depression-with-tdcs-startup-ybrain-aims-for-the-mainstream.

“Students – Facts & Figures – About WU.” Accessed September 23, 2025. https://www.wu.ac.at/en/the-university/about-wu/facts-figures/studierende.

“Studien.” January 2, 2025. https://www.studieren.at/studien/.

Süddeutsche Zeitung. “Kräuter Statt Krypto: Krisenfestes Investment in Die Zukunft.” June 4, 2025. https://pressemitteilungen.sueddeutsche.de/sonnentor-krauterhandelsgesellschaft-mbh-6048822.

Sunstein, Cass R. *Republic.Com 2.0*. Princeton University Press, 2007.

Taatgen, Niels A., and Lee, Frank J. Production Compilation: A Simple Mechanism to Model Complex Skill Acquisition. *Human Factors* 45, no. 1 (2003): 61–76. https://doi.org/10.1518/hfes.45.1.61.27224.

Taylor, Charles. *A Secular Age*. Belknap Press, 2007.

Testa, Stefania, Forsman, Helena, Mattarelli, Elisa, and Petruzzelli, Antonio Messeni. Innovation Failure: What We Know and What Is Missing. *Technovation* 140, (February 2025): 103140. https://doi.org/10.1016/j.technovation.2024.103140.

Thain, Marion. Distracted Reading: Acts of Attention in the Age of the Internet. *Digital Humanities Quarterly* 012, no. 2 (2018).

Thaler, Richard H., and Sunstein, Cass R. *Nudge: Improving Decisions about Health, Wealth, and Happiness*. Yale University Press, 2008.

The National. “Artificial Muscles Set to Create Robots with Super Powers.” November 29, 2017. https://www.thenationalnews.com/business/technology/artificial-muscles-set-to-create-robots-with-super-powers-1.679896.

Thimbleby, Harold. *Fix IT: See and Solve the Problems of Digital Healthcare*. Oxford University Press, 2021. https://doi.org/10.1093/oso/9780198861270.001.0001.

Thompson, Karl. “What Percentage of Your Life Will You Spend at Work? – ReviseSociology.” August 16, 2016. https://revisesociology.com/2016/08/16/percentage-life-work/.

Tobin, Donal. “Data Transformation Challenge Statistics – 50 Statistics Every Technology Leader Should Know in 2025.” Integrate.Io, August 24, 2025. https://www.integrate.io/blog/data-transformation-challenge-statistics/.

Třebický, Vít, Delplanque, Sylvain, Ferdenzi, Camille, et al. Cross-Modal Associations of Human Body Odour Attractiveness with Facial and Vocal Attractiveness Provide Little Support for the Backup Signals Hypothesis: A Systematic Review and Meta-Analysis. *Evolution and Human Behavior* 44, no. 1 (2023): 19–29. https://doi.org/10.1016/j.evolhumbehav.2022.11.001.

Trojanow, Ilija. *Der überflüssige Mensch: Unruhe bewahren*. Residenz Verlag, 2013.

Turkle, Sherry. Authenticity in the Age of Digital Companions. *Interaction Studies* 8, no. 3 (2007): 501–517. https://doi.org/10.1075/is.8.3.11tur.

Turkle, Sherry, Taggart, Will, Kidd, Cory D., and Dasté, Olivia. Relational Artifacts with Children and Elders: The Complexities of Cybercompanionship. *Connection Science* 18, no. 4 (2006): 347–361. https://doi.org/10.1080/09540090600868912.

Tyng, Chai M., Amin, Hafeez U., Saad, Mohamad N. M., and Malik, Aamir S. The Influences of Emotion on Learning and Memory. *Frontiers in Psychology* 8, (August 2017): 1454. https://doi.org/10.3389/fpsyg.2017.01454.

Ulrich, Sarah. "»Ohne uns würde es nicht funktionieren« Theresa Ingendaay über die Gründung der Fahrradkurier-Gewerkschaft." Kreuzer Online, December 6, 2017. https://kreuzer-leipzig.de/2017/12/06/ohne-uns-wuerde-es-nicht-funktionieren.

"Uptake of Open Access." *STM Association*, n.d. Accessed December 4, 2025. https://stm-assoc.org/oa-dashboard/oa-dashboard-2/uptake-of-open-access/.

Vallor, Shannon. *Technology and the Virtues: A Philosophical Guide to A Future Worth Wanting*. Oxford University Press, 2016.

Van Noorden, Richard. How Big Is Science's Fake-Paper Problem?. *Nature* 623, no. 7987 (2023): 466–467. https://doi.org/10.1038/d41586-023-03464-x.

Villeneuve, Crispian. *Rudolf Steiner: The British Connection : Elements from His Early Life and Cultural Development*. Temple Lodge Publishing, 2011.

Virilio, Paul. *Polar Inertia*. 1st ed. SAGE Publications Ltd, 1999.

Vrselja, Zvonimir, Daniele, Stefano G., Silbereis, John, et al. Restoration of Brain Circulation and Cellular Functions Hours Post-Mortem. *Nature* 568, no. 7752 (2019): 336–343. https://doi.org/10.1038/s41586-019-1099-1.

Wakabayashi, Daisuke, and Shane, Scott. "Google Will Not Renew Pentagon Contract That Upset Employees." Technology. *The New York Times*, June 1, 2018. https://www.nytimes.com/2018/06/01/technology/google-pentagon-project-maven.html.

Wansink, Brian, Painter, James E., and North, Jill. Bottomless Bowls: Why Visual Cues of Portion Size May Influence Intake. *Obesity Research* 13, no. 1 (2005): 93–100. https://doi.org/10.1038/oby.2005.12.

Warren, Samuel D., and Brandeis, Louis D. The Right to Privacy. *Harvard Law Review* 4, no. 5 (1890): 193–220. https://doi.org/10.2307/1321160.

Webb, Michael, Jones, Chad, Van Reenen, John, and Nicholas, Bloom. "Ideas Aren't Running Out, but They are Getting More Expensive to Find." VOX CEPR. *VOXEU Productivity and Innovation*, September 20, 2017. https://cepr.org/voxeu/columns/ideas-arent-running-out-they-are-getting-more-expensive-find.

Weber, Max. *The Protestant Ethic and the Spirit of Capitalism*. Translated by Talcott Parsons. Dover Publications, 2012.

Weil, Simone. *Gravity and Grace*. Translated by Emma Craford and Mario von der Ruhr. Routledge Classics, 2002.

Weinberger, David. "The Problem with the Data-Information-Knowledge-Wisdom Hierarchy." *Harvard Business Review*, 2010. https://hbr.org/2010/02/data-is-to-info-as-info-is-not, https://hbr.org/2010/02/data-is-to-info-as-info-is-not.

Welle, Elissa. "Aligning Those Who Align AI, One Satirical Website at a Time." *The Verge*, September 11, 2025. https://www.theverge.com/ai-artificial-intelligence/776752/center-for-the-alignment-of-ai-alignment-centers.

Weller, Chris. "Bill Gates and Steve Jobs Raised Their Kids Tech-Free – And It Should've Been a Red Flag." *The Independent*, April 11, 2018. https://www.independent.co.uk/tech/bill-gates-and-steve-jobs-raised-their-kids-techfree-and-it-should-ve-been-a-red-flag-a8017136.html.

Wheelwright, Trevor. "Cell Phone Usage Stats 2025: Americans Check Their Phones 205 Times a Day." *Reviews.Org*, January 1, 2025. https://www.reviews.org/mobile/cell-phone-addiction/.

WIPO IP Facts and Figures 2024. World Intellectual Property Organization, 2024. https://www.wipo.int/web-publications/ip-facts-and-figures-2024/assets/70866/943EN_IP%20Facts%20and%20Figures%202024.4.pdf.

wko.at. "Lehrlingsstatistik." Accessed December 4, 2025. https://www.wko.at/zahlen-daten-fakten/daten-lehrlingsstatistik.

"Wolfie Christl." Accessed December 4, 2025. https://wolfie.crackedlabs.org/en.

"Word Happiness Report." August 20, 2025. https://worldhappiness.report/.

Wundt, Wilhelm. *Logik*. Vol. 1. Veit & Comp, 1919.

Wurzman, Rachel, Hamilton, Roy H., Pascual-Leone, Alvaro, and Fox, Michael D. An Open Letter Concerning Do-It-Yourself Users of Transcranial Direct Current Stimulation. *Annals of Neurology* 80, no. 1 (2016): 1–4. https://doi.org/10.1002/ana.24689.

Youth Perspectives on Online Safety, 2024: An Annual Report of Youth Attitudes and Experiences. Thorn, 2025. https://info.thorn.org/hubfs/Research/Thorn_24_YouthMonitoring_Report.pdf.

Youyou, Wu, Kosinski, Michal, and Stillwell, David. Computer-Based Personality Judgments are More Accurate than Those Made by Humans. *Proceedings of the National Academy of Sciences of the United States of America (PNAS)* 112, no. 4 (2015): 1036–1040.

Zeiler, Michael D. Eliminating Behavior with Reinforcement. *Journal of the Experimental Analysis of Behavior* 16, no. 3 (1971): 401–405. https://doi.org/10.1901/jeab.1971.16-401.

Zuboff, Shoshana. *The Age of Surveillance Capitalism: The Fight for a Human Future at the New Frontier of Power*. Profile Books, 2019.

Endnotes

1 I am deliberately using the term "value creation" here in a very different sense than is used in modern economics, where a GDP and monetary focus prevail that I do not believe in.
2 Kurzweil, Ray. 2006. *The Singularity is Near: When Humans Transcend Biology*. London: Penguin Group.
3 Classical relics often include teeth, skulls, scraps of fabric or similar items that belonged to the person who is worshipped as part of the cult.
4 (German edition of) Kurzweil, Ray. 2006. *The Singularity Is Near: When Humans Transcend Biology*. London: Penguin Group.
5 Kohl, Karl-Heinz. 2003. *Die Macht der Dinge – Geschichte und Theorie sakraler Objekte*. München: C. H. Beck (own translation).
6 Belk, Russell and Gülnur and Tumbat, "The Cult of Macintosh," Consumption Markets & Culture, no. 3 (2005): 205–17, https://doi.org/10.1080/10253860500160403.
7 "Is AI Conscious? Former Google Employee Speaks Out." https://www.youtube.com/watch?v=yk35jJWuKtI
8 See Jaeger, Johannes, Anna Riedl, Alex Djedovic, John Vervaeke, and Denis Walsh. 2025. "Naturalizing relevance realization: why agency and cognition are fundamentally not computational." *Frontiers in Psychology* 15. doi: https://doi.org/10.3389/fpsyg.2024.1362658.
9 See Khamassi, Mehdi, Marceau Nahon, and Raja Chatila. 2024. "Strong and weak alignment of large language models with human values." *Nature – Scientific Reports* 14. doi: https://doi.org/10.1038/s41598-024-70031-3.
10 I also neglect the many so-called narrow AI systems that have been helping us improve functionality in our devices and machines for decades (such as the quality of our cell phone cameras, navigation systems, etc.).
11 "All Too Human? Mapping and Mitigating the Risks from Anthropomorphic AI." Akbulut, Canfer, Laura Weidinger, Arianna Manzini, Iason Gabriel, Verena Rieser Google DeepMind, Proceedings of the Seventh AAAI/ACM Conference on AI, Ethics, and Society (AIES 2024).
12 "Japanerin heiratet ihren Traumpartner, es ist ChatGPT," 17.12.2025 (last visited 2.1.2026). https://www.youtube.com/watch?v=KveRPtRq75k
13 BBC; "A predator in your home: Mothers say that chatbots encouraged their sons to kill themselves." 4. November 2025; URL: https://www.bbc.com/news/articles/ce3xgwyywe4o
14 Spiekermann, Sarah, Marc Langheinrich, Johannes Hoff, Christiane Wendehorst, Jürgen Pfeffer, Thomas Fuchs, and Armin Grunwald. 2025. "The Power of 10: New Rules for the Digital World." Communication of the ACM forthcoming. doi: http://dx.doi.org/10.1145/3774658.
15 Istrate, Robert, Victor Tulus, Robert N. Grass, Laurent Vanbever, Wendelin J. Stark, and Gonzalo Guillén-Gosálbez. 2024. "The environmental sustainability of digital content consumption." *Nature Communications* (15).
16 International Association of Electronic Waste Producer Responsibility Organisations, https://weee-forum.org/ (3.1.2025).
17 Crawford, Kate. 2021. *Atlas of AI – Power, Politics, and the Planetary Costs of Artificial Intelligence*: Yale University Press.
18 Qiao, Biting. 2025, Nontarget Screening and Occurrence of Emerging Per- and Polyfluoroalkyl Substances in Municipal and Semiconductor Industrial Wastewater: A Large-Scale Survey in China; *Environmental Science & Technology*; 59, no. 23. https://pubs.acs.org/doi/abs/10.1021/acs.est.5c02035

 | https://doi.org/10.1515/9783111677736-011

19 International Association of Electronic Waste Producer Responsibility Organisations, https://weee-forum.org/ (3.1.2025).
20 For legal reasons, the actual name of the company was changed.
21 Redman, Thomas C.: "Data quality problems plague every department, in every industry, at every level, and for every type of information studies show that knowledge workers waste up to 50% of time hunting for data, identifying and correcting errors, and seeking confirmatory sources for data they do not trust" in: Thomas C. Redman, "Data's Credibility Problem," *Harvard Business Review*, 2013, 2. And for an overview of recent surveys see Donal Tobin, "Data Transformation Challenge Statistics – 50 Statistics Every Technology Leader Should Know in 2025," Integrate. Io, August 24, 2025, https://www.integrate.io/blog/data-transformation-challenge-statistics/.
22 Examples include Ovum and Gartner, but also many smaller providers.
23 "[V]alues are irreducible basic phenomena of emotive intuition" Max Scheler, *Formalism in Ethics and Non-Formal Ethics of Values: A New Attempt toward the Foundation of an Ethical Personalism*, trans. Manfred S. Frings and Roger L. Funk (Northwestern University Press, 1973), 265, http://archive.org/details/formalisminethic0000sche.
24 The idea of negative values that is discussed in Max Scheler's value ethics is disputed. Can there even be negative values if what is valuable is always something that you value? The inversion seems problematic. Nevertheless I use the concept in this book mainly for the sake of simplicity.
25 Hanna Krasnova et al., "Why Following Friends Can Hurt You: An Exploratory Investigation of the Effects of Envy on Social Networking Sites among College-Age Users," *Information Systems Research* (US) 26, no. 3 (2015): 585–605, https://doi.org/10.1287/isre.2015.0588. and Fabio Carraturo et al., "Envy, Social Comparison, and Depression on Social Networking Sites: A Systematic Review," *European Journal of Investigation in Health, Psychology and Education* 13, no. 2 (2023): 364–76, https://doi.org/10.3390/ejihpe13020027.
26 Nir Eyal, *Hooked: How to Build Habit-Forming Products* (Penguin Canada, 2014).
27 Picture is based on "Gartner 2024 Hype Cycle for Emerging Technologies Highlights Developer Productivity, Total Experience, AI and Security," Gartner, August 21, 2024, https://www.gartner.com/en/newsroom/press-releases/2024-08-21-gartner-2024-hype-cycle-for-emerging-technologies-highlights-developer-productivity-total-experience-ai-and-security.
28 Mariana Mazzucato, *The Value of Everything: Making and Taking in the Global Economy* (Allen Lane, 2018).
29 "World Happiness Report," August 20, 2025, https://worldhappiness.report/.
30 Albert Borgmann, *Technology and the Character of Contemporary Life: A Philosophical Inquiry* (University of Chicago Press, 1984).
31 Quotation: "By subjecting technological innovation completely to the logic of the market, the so-called conservative revolution led by Thatcher and Reagan has engendered a cultural and spiritual regression of unprecedented magnitude, transforming the whole of society into a machine for profit maximization and creating a state of 'systemic carelessness' and 'systemic stupidity' on a global scale." Pieter Lemmens, "'This System Does Not Produce Pleasure Anymore'. An Interview with Bernard Stiegler," *Krisis | Journal for Contemporary Philosophy* 31, no. 1 (2011): 33–41.
32 Ilija Trojanow, *Der überflüssige Mensch: Unruhe bewahren* (Residenz Verlag, 2013).
33 Interview with Foodora couriers conducted by A. Reinsberg und C. Weiske Alexandra Reinsberg and Cosima Weiske, *Interview mit einem Foodora-Lieferanten: "Es ist ein Knochenjob,"* November 8, 2016, https://www.br.de/puls/themen/leben/arbeiten-fuer-foodora-so-ist-es-wirklich-interview-100.html.

34 Sarah Ulrich, "»Ohne uns würde es nicht funktionieren« Theresa Ingendaay über die Gründung der Fahrradkurier-Gewerkschaft," *Kreuzer Online*, December 6, 2017, https://kreuzer-leipzig.de/2017/12/06/ohne-uns-wuerde-es-nicht-funktionieren.
35 Balachandar Aiswarya and Gurusamy Ramasundaram, "Online Food Delivery Technology and Family Food Traditions: A Sociocultural Perspective," *Journal of Information Technology Teaching Cases* 14, no. 1 (2024): 68–73, https://doi.org/10.1177/20438869221150826.
36 Lilly Irani, "Difference and Dependence among Digital Workers: The Case of Amazon Mechanical Turk," *South Atlantic Quarterly* 114, no. 1 (2015): 225–34, https://doi.org/10.1215/00382876-2831665.
37 With the classical roadmap analysis there were 139 product ideas; with the ethically inspired analysis there were 453.
38 Scheler, *Formalism in Ethics and Non-Formal Ethics of Values*, 16.
39 Scheler, *Formalism in Ethics and Non-Formal Ethics of Values*, 265.
40 Scheler speaks of the "ultimate independence of the being of values with regard to things, goods, and states of affairs" Scheler, *Formalism in Ethics and Non-Formal Ethics of Values*, 17.
41 Saffron Huang et al., "Values in the Wild: Discovering and Analyzing Values in Real-World Language Model Interactions," arXiv:2504.15236, preprint, arXiv, April 21, 2025, https://doi.org/10.48550/arXiv.2504.15236.
42 Liangkang Ni, *Zur Sache des Bewusstseins: Phänomenologie – Buddhismus – Konfuzianismus* (Königshausen u. Neumann, 2010).
43 Scheler, *Formalism in Ethics and Non-Formal Ethics of Values.*
44 Scheler developed phenomenology further. See Eugene Kelly, *Material Ethics of Value: Max Scheler and Nicolai Hartmann*, vol. 203, Phaenomenologica (Springer Netherlands, 2011), https://doi.org/10.1007/978-94-007-1845-6.
45 Scheler, *Formalism in Ethics and Non-Formal Ethics of Values*, 35.
46 Many people today are, on the other hand, quite comfortable with the idea that striving towards something is the key to the good. The careful consideration of our true desires, however, should not be underestimated. We must not confuse the striving for values with the striving for something we want, with a want for self-conceived ideas about life. As Scheler writes: "Goals of conation differ most distinctly from purposes of the will. The goal lies in the very process of the conation. It is not conditioned by some act of representation, but is as immanent to the conation as 'content' is to representation." Scheler, *Formalism in Ethics and Non-Formal Ethics of Values*, 34–35.
47 Scheler, *Formalism in Ethics and Non-Formal Ethics of Values*, 42.
48 Scheler, *Formalism in Ethics and Non-Formal Ethics of Values*, 147.
49 The academic literature uses the terminology of extrinsic values that are instrumental for intrinsic values. Intrinsic values are values in themselves. These can easily be recognized as no one calls their importance into question because they appear desirable in themselves.
50 These requirements or dispositions can also be described as the "features" of an apartment.
51 Scheler, *Formalism in Ethics and Non-Formal Ethics of Values*, 20.
52 Scheler, *Formalism in Ethics and Non-Formal Ethics of Values*, 266. I am aware that some cognitive psychologists do not share this view of world perception. They believe that the world only exists because we perceive it. I object to this view – in fact, I consider this academic view untenable and outdated.
53 Sarah Spiekermann, *Ethical IT Innovation: A Value-Based System Design Approach* (Taylor & Francis, 2016), 136–39.

54 Ikujirō Nonaka et al., eds., *Managing Flow: A Process Theory of the Knowledge-Based Firm* (Palgrave Macmillan, 2008).
55 Scheler, *Formalism in Ethics and Non-Formal Ethics of Values*, p. 85.
56 Aristotle, *Aristotle's Nicomachean Ethics*, trans. Robert C Bartlett and Susan D. Collins (University of Chicago Press, 2011), http://archive.org/details/aristotlesnicoma0000aris.
57 G. E. M. Anscombe, "Modern Moral Philosophy," *Philosophy* 33, no. 124 (1958): 1–19.
58 Alasdair MacIntyre, *After Virtue: A Study in Moral Theory* (University of Notre Dame Press, 1984).
59 Sandel, Michael J. *Justice: What's the Right Thing to Do?* Penguin UK, 2009.
60 Shannon Vallor, *Technology and the Virtues: A Philosophical Guide to a Future Worth Wanting* (Oxford University Press, 2016), 63.
61 Doug Henton and Kim Held, "The Dynamics of Silicon Valley: Creative Destruction and the Evolution of the Innovation Habitat," *Social Science Information* 52, no. 4 (2013): 539–57, https://doi.org/10.1177/0539018413497542.
62 Michael E. Porter and Mark R. Kramer, "Creating Shared Value," Corporate Social Responsibility, *Harvard Business Review*, January 1, 2011, https://hbr.org/2011/01/the-big-idea-creating-shared-value.
63 Ikujiro Nonaka and Hirotaka Takeuchi, "The Wise Leader," *Harvard Business Review*, 2011, 61.
64 Max Scheler's *Formalism in Ethics and Non-formal Ethics of Values* was taken up by Nicolai Hartmann in his axiological value ethics. Both philosophically rigorous value philosophies have, however, been overshadowed by almost superficially simple lists of values that have developed from psychology, which often lack fundamental substance and have instead only created confusing discussions around value pluralism.
65 *Alexander Osterwalder – Value Proposition Design*, produced by FBTB, 2017, 58:51, https://www.youtube.com/watch?v=b_X18bmpHaw.
66 Walter Isaacson, *Steve Jobs* (Simon & Schuster, 2011), 133.
67 Michael E. Porter, *Competitive Strategy: Techniques for Analyzing Industries and Competitors* (Free Press, 2008), chap. 2.
68 Evelyn Cheng, "Just 10% of Trading Is Regular Stock Picking, JPMorgan Estimates," Markets, *CNBC*, June 13, 2017, https://www.cnbc.com/2017/06/13/death-of-the-human-investor-just-10-percent-of-trading-is-regular-stock-picking-jpmorgan-estimates.html.
69 Isaacson, *Steve Jobs*, 157.
70 This figure was cited by a VW works council member during an internal discussion I attended with Internet 4.0 experts at the German Trade Union Confederation (Deutschen Gewerkschaftsbund) in 2017.
71 Sebastian Herrera, "Exclusive | Amazon Is on the Cusp of Using More Robots Than Humans in Its Warehouses," *The Wall Street Journal*, accessed September 10, 2025, https://www.wsj.com/tech/amazon-warehouse-robots-automation-942b814f.
72 Sheelah Kolhatkar, "Welcoming Our New Robot Overlords," Annals of Technology, *The New Yorker*, October 16, 2017, https://www.newyorker.com/magazine/2017/10/23/welcoming-our-new-robot-overlords.
73 Heather Haddon, "Rise of the Restaurant Robots: Chipotle, Sweetgreen and Others Bet on Automation," *The Wall Street Journal*, July 11, 2024, https://www.wsj.com/business/hospitality/restaurant-robots-flippy-chippy-autocado-9de44eeb.https://web.archive.org/web/20241010091437/https://www.wsj.com/business/hospitality/restaurant-robots-flippy-chippy-autocado-9de44eeb.

74 "Table 1. Median Years of Tenure with Current Employer for Employed Wage and Salary Workers by Age and Sex, Selected Years, 2014–2024 – 2023 A01 Results," Bureau of Labor Statistics, accessed November 24, 2025, https://www.bls.gov/news.release/tenure.t01.htm.
75 "SONNENTOR Organic Teas & Organic Spices," accessed September 10, 2025, https://www.sonnentor.com/en-gb/about-us/history.
76 James Shotter, "A Taste for Risk and Something Different," *Financial Times*, November 20, 2012, https://www.ft.com/content/31a2f6cc-3243-11e2-b891-00144feabdc0.
77 "Kräuter Statt Krypto: Krisenfestes Investment in Die Zukunft," *Süddeutsche Zeitung*, June 4, 2025, https://pressemitteilungen.sueddeutsche.de/sonnentor-krauterhandelsgesellschaft-mbh-6048822.
78 "Österreich – Beste Gewürzmarken 2024," Statista, accessed December 3, 2025, https://de.statista.com/statistik/daten/studie/1405676/umfrage/ranking-der-besten-marken-fuer-gewuerze-in-oesterreich/.
79 In 2023, the average hourly wage of a microworker was 2.15 USD/hour: "Fairwork Cloudwork Ratings 2023: Work in the Planetary Labour Market," accessed September 10, 2025, https://fair.work/en/fw/publications/fairwork-cloudwork-ratings-2023-work-in-the-planetary-labour-market/. See also Alana Semuels, "The Internet Is Enabling a New Kind of Poorly Paid Hell," Business, *The Atlantic*, January 23, 2018, https://www.theatlantic.com/business/archive/2018/01/amazon-mechanical-turk/551192/. This article cites the study: Kotaro Hara et al., "A Data-Driven Analysis of Workers' Earnings on Amazon Mechanical Turk," *Proceedings of the 2018 CHI Conference on Human Factors in Computing Systems* (New York, NY, USA), CHI '18, Association for Computing Machinery, April 21, 2018, 1–14, https://doi.org/10.1145/3173574.3174023.
80 My knowledge is based on a workshop at the Vienna University of Economics and Business in September 2018, where a former Foodora courier and current researcher in the field of micro-jobs gave a detailed account of the conditions. "Dein Chef ist eine App", Privacy & Sustainable Computing Lab, WU Vienna.
81 The case description is taken from the *Aeon* digital magazine. Vincent Gabrielle, "How Employers Have Gamified Work for Maximum Profit | *Aeon* Essays," *Aeon*, October 10, 2018, https://aeon.co/essays/how-employers-have-gamified-work-for-maximum-profit.
82 Shona Ghosh, "Undercover Author Finds Amazon Warehouse Workers in UK 'peed in Bottles' over Fears of Being Punished for Taking a Break," *Business Insider*, accessed September 10, 2025, https://www.businessinsider.com/amazon-warehouse-workers-have-to-pee-into-bottles-2018-4.
83 Scheler, *Formalism in Ethics and Non-Formal Ethics of Values*, 86–104.
84 "Nevertheless, this order of ranks and values is a *non-formal* order, an order of *value-qualities*. And insofar as such an order of values is not absolute but 'predominant,' it is represented in the rules of preferring among value-qualities which embody a given epoch. In the sphere of aesthetic values we call the systems of such rules of preferring a 'style,' and in the sphere of practical values, a 'morality.' Such systems reveal growth and development. But this development is *totally* different from that of the world of goods and is a variable independent of it." (Emphasis in original). Scheler, *Formalism in Ethics and Non-Formal Ethics of Values*, 23.
85 Amazon Turk is one of the oldest platforms. See "Amazon Turk," n.d., accessed September 10, 2025, https://www.mturk.com/.See: https://www.mturk.com/ (accessed 10/09/2025).
86 Felicity Lawrence, "Uber Is Treating Its Drivers as Sweated Labour, Says Report," Technology, *The Guardian*, December 9, 2016, https://www.theguardian.com/technology/2016/dec/09/uber-drivers-report-sweated-labour-minimum-wage.
87 When I use the word "invaluable" below, I mean "saturated with values."

88 Artificial intelligence works with probabilities. It observes or detects patterns based on the likelihood with which something occurs. This means it is hardly able to recognize the unique details that give meaning to each individual human being.
89 The concept of "singularitas" is, in this book, based on historical and philosophical sources. In particular, as it was characterized by Nicholas of Cusa (1401-1464). Cf. Gerda Bredow, *Participatio Singularitatis. Einzigartigkeit als Grundmuster der Weltgestaltung*, Archiv Für Geschichte Der Philosophie, 71, no. 2 (1989): 216–30, https://doi.org/10.1515/agph.1989.71.2.216. See also Johannes Hoff, *The Analogical Turn: Rethinking Modernity with Nicholas of Cusa* (William B. Erdmans Publishing Company, 2013), 153–67. The concept of "singularitas" that I use here should not be confused with the new pop-concept of singularity as characterized by Ray Kurzweil in his reflections on the future. Kurzweil defines the singularity as the moment at which artificial intelligence exceeds human intelligence. However, his book neglects the historical sources on which the term is based. Kurzweil mainly extrapolates projections from computer history and adopts the metaphor of singularity in physics to construct his own worldview. In this respect, it seems to me that he is completely unfamiliar with the historical depth of singularitas.
90 For a critical discussion, see also David Weinberger, "The Problem with the Data-Information -Knowledge-Wisdom Hierarchy," *Harvard Business Review*, 2010, https://hbr.org/2010/02/data-is-to -info-as-info-is-not, https://hbr.org/2010/02/data-is-to-info-as-info-is-not.
91 "Dataism" could also be an allusion by Harari to the movement of "Dadaism" in art, which was, above all, characterized by its rejection of "conventional" art and art forms and created highly surreal images of reality.
92 See Jeana H. Frost et al., "People Are Experience Goods: Improving Online Dating with Virtual Dates," *Journal of Interactive Marketing* 22, no. 1 (2008): 51–61, https://doi.org/10.1002/dir.20107., and: Bettina Arndt, "What Gives a First Online Date Every Chance of Turning into a Second?," The Sydney Morning Herald, November 16, 2016, https://www.smh.com.au/lifestyle/life-and-relationships/what-gives-a-first-online-date-every-chance-of-turning-into-a-second-20161116-gsqg12.html.
93 The sending of fragrance samples over the Internet is already possible: Euronews, "Share Touch, Smell and Taste via the Internet," 2015. The only thing left to do is to effectively capture your own daily scent.
94 Vít Třebický et al., "Cross-Modal Associations of Human Body Odour Attractiveness with Facial and Vocal Attractiveness Provide Little Support for the Backup Signals Hypothesis: A Systematic Review and Meta-Analysis," *Evolution and Human Behavior* 44, no. 1 (2023): 19–29, https://doi.org/10.1016/j.evolhumbehav.2022.11.001.
95 See, for example, the Samsung ES8000er Smart-TVs and their media coverage: "Is Your TV Watching You? Latest Models Raise Concerns," NBC News, March 19, 2012, http://www.nbcnews.com/technolog/your-tv-watching-you-latest-models-raise-concerns-483619.
96 Our personal data is today shared and sold very efficiently on global networked data markets. Regarding the company Lenddo see Tobias Berg et al., "On the Rise of FinTechs: Credit Scoring Using Digital Footprints," *The Review of Financial Studies* 33, no. 7 (2020): 2845–97. How these and other mechanisms precisely function is lucidly described in: W. Christl, "Corporate Surveillance in Everyday Life. How Companies Collect, Combine, Analyze, Trade, and Use Personal Data on Billions," 2017., Described in Wolfie Christl and Sarah Spiekermann, *Networks of Control: A Report on Corporate Surveillance, Digital Tracking, Big Data & Privacy* (Facultas, 2016).
97 The dating app Score, launched in 2024, requires, for instance, a high credit score: Claer Barrett, "A Dating App with Credit Scores: What's Not to Love? Proving You Can Manage Your Finan-

ces Has Real Pulling Power," *Financial Times*, February 10, 2024, https://www.ft.com/content/7ad0c6f5-c3bf-4648-94ad-f3bebccbdf87.

98 On the idea of mirrors, see Evgeny Morozov, *To Save Everything, Click Here* (Penguin, 2013).

99 Christoph Engel et al., "Code Is Law: How COMPAS Affects the Way the Judiciary Handles the Risk of Recidivism," *Artificial Intelligence and Law* 33, no. 2 (2025): 383–404, https://doi.org/10.1007/s10506-024-09389-8.

100 Methodologically dubious top-tier journal articles claim "scientific" evidence to support this claim. For example: W. Youyou et al., "Computer-Based Personality Judgments Are More Accurate than Those Made by Humans," *Proceedings of the National Academy of Sciences of the United States of America (PNAS)* 112, no. 4 (2015): 1036–40.

101 Psychologists talk about "psychic entropy". See Mihaly Csikszentmihalyi and Isabella Selega Csikszentmihalyi, *Optimal Experience: Psychological Studies of Flow in Consciousness* (University Press, 1988), 22.

102 See M. I. Norton, "Less Is More: The Lure of Ambiguity or Why Familiarity Breeds Contempt," *Journal of Personality and Social Psychology* 92, no. 1 (2007): 97–105.

103 At this point, we can critically compare the alternative of not spending seven hours exchanging e-mails and text messages with the other person, but instead spending seven hours talking. If 100 potential couples (who were selected for each other via a dating platform) were placed in such a situation, would 90 couples assume that, after these seven hours, they have so much in common that they would arrange a second date? In fact, statistics tell us that only 20–30% (rather than 90%) can say that the saturated impression of their first meeting was so successful that they would arrange a second date.

104 In the USA, 42% of online daters report that this form has made dating easier. See Colleen McClain and Risa Gelles-Watnick, "From Looking for Love to Swiping the Field: Online Dating in the U.S.," *Pew Research Center*, February 2, 2023, https://www.pewresearch.org/internet/2023/02/02/from-looking-for-love-to-swiping-the-field-online-dating-in-the-u-s/.

105 Psychologist Ellen Langer even shows that control over things correlates highly with human lifespan. Ellen Langer, *The Psychology of Control* (Sage Publications, 1983), 310.

106 James R. Averill, "Personal Control over Aversive Stimuli and Its Relation to Stress," *Psychological Bulletin* 80 (1973): 286–303.

107 In psychology, the reaction of sudden anger as a result of loss of control is called "reactance." See Jack W. Brehm, *A Theory of Psychological Reactance* (Academic Press, 1966); Sharon S. Brehm and Jack W. Brehm, *Psychological Reactance: A Theory of Freedom and Control* (Academic Press, 1981). In relation to IT: Steven M. Edwards et al., "Forced Exposure and Psychological Reactance: Antecedents and Consequences of the Perceived Intrusiveness of Pop-Up Ads," *Journal of Advertising* 31, no. 3 (2002): 83–95, https://doi.org/10.1080/00913367.2002.10673678.

108 L. Y. Abramson et al., "Learned Helplessness in Humans," *Journal of Abnormal Psychology* 87, no. 1 (1978): 49–74. In the context of IT: Oliver Guenther and Sarah Spiekermann, "RFID and Perceived Control: The Consumer's View," *Communications of the ACM* 48, no. 9 (2005): 73–76.

109 In chemistry class today, we still learn that our biological bodies are also divided in this way. We have developed many good models of how we can break ourselves down to the atomic level; at the subatomic level into protons, neutrons, and electrons. However, even these are not the elementary particles capable of splitting our unity, because quarks are even smaller. Quarks, in turn, cannot be split so easily because they always form a unit with other quarks; they combine to form so-called hadrons. So, to cut a long story short: the argument often put forward that humans are also made up of tiny particles is scientifically unproven and rather unlikely.

110 *Wikipedia*, “Turing test,” September 11, 2025, https://en.wikipedia.org/w/index.php?title=Turing_test&oldid=1310747226.
111 “Facebook Fake Account Deletion per Quarter 2024,” 2025.
112 This estimate comes from: James Somers, “The Coming Software Apocalypse,” Technology, *The Atlantic*, September 26, 2017, https://www.theatlantic.com/technology/archive/2017/09/saving-the-world-from-code/540393/. However, thanks to modern and more secure programming methods such as “correct by construction,” this situation is changing. The latter, however, is only used in very select industries. See Tobias Runge et al., “Comparing Correctness-by-Construction with Post-Hoc Verification – A Qualitative User Study,” in *Formal Methods. FM 2019 International Workshops*, ed. Emil Sekerinski et al. (Springer International Publishing, 2020), https://doi.org/10.1007/978-3-030-54997-8_25.
113 Mark Sherman, “Using ChatGPT to Analyze Your Code? Not So Fast,” February 12, 2024, https://insights.sei.cmu.edu/blog/using-chatgpt-to-analyze-your-code-not-so-fast/.
114 The use of machine learning on erroneous code increases the risk of reusing error-prone code. William Harding, *AI Copilot Code Quality* (GitClear, 2025), https://gitclear-public.s3.us-west-2.amazonaws.com/GitClear-AI-Copilot-Code-Quality-2025.pdf.
115 James Somers, “The Coming Software Apocalypse,” Technology, The Atlantic, September 26, 2017.
116 James Somers, “The Coming Software Apocalypse,” Technology, The Atlantic, September 26, 2017.
117 For a recent report see Sharon Barbour et al., “NHS Computer Problems Put Patients at Risk of Harm,” *BBC News*, May 30, 2024, https://www.bbc.com/news/articles/c4nn0vl2e78o. For a more in-depth study: Harold Thimbleby, *Fix IT: See and Solve the Problems of Digital Healthcare* (Oxford University Press, 2021), https://doi.org/10.1093/oso/9780198861270.001.0001.
118 The numbers vary: Wikipedia reports 26,000 parents (*Wikipedia*, “Dutch childcare benefits scandal,” July 31, 2025, https://en.wikipedia.org/w/index.php?title=Dutch_childcare_benefits_scandal&oldid=1303491793) and other sources say that over 70,000 people registered for compensation (Robin Pascoe, “Over 33,000 Families Acknowledged as Benefit Scandal Victims,” *DutchNews.Nl*, January 22, 2024, https://www.dutchnews.nl/2024/01/over-33000-families-acknowledged-as-benefit-scandal-victims/).
119 Hanneke Sanou, “Over 1,100 Children of Childcare Benefit Victims Taken into Care,” *DutchNews.Nl*, October 20, 2021, https://www.dutchnews.nl/2021/10/over-1100-children-of-childcare-benefit-victims-taken-into-care/.
120 “EXPLANATORY NOTES Post Office (Horizon System) Offences Act 2024 Chapter 14,” n.d., 3, accessed December 3, 2025, https://www.legislation.gov.uk/ukpga/2024/14/pdfs/ukpgaen_20240014_en.pdf.
121 This is because the explanation itself is a model of what developers presume the black box to do. Will Knight, “The Dark Secret at the Heart of AI,” *MIT Technology Review* 120, no. 3 (2017): 54–63.
122 Sarah Spiekermann et al., “Inside the Organization: Why Privacy and Security Engineering Is a Challenge for Engineers,” *Proceedings of the IEEE* 107, no. 3 (2019): 600–615, https://doi.org/10.1109/JPROC.2018.2866769.
123 Brian Berenbach and Manfred Broy, “Professional and Ethical Dilemmas in Software Engineering,” *IEEE Computer*, no. 1 (2009): 74–80, https://doi.org/10.1109/MC.2009.22.
124 Here is, for example, the IEEE Standard for the Transparency of Autonomous Systems (IEEE 7001–2021). See “IEEE Standard for the Transparency of Autonomous Systems (IEEE 7001–2021),” 2021.
125 Scheler, *Formalism in Ethics and Non-Formal Ethics of Values*, 85.

126 "Time Spent Using Smartphones (2025 Statistics)," Exploding Topics, March 28, 2025, https://explodingtopics.com/blog/smartphone-usage-stats.
127 "Worldwide Daily Social Media Usage (New 2025 Data)," Exploding Topics, May 4, 2022, https://explodingtopics.com/blog/social-media-usage.
128 The digital presents itself to us as the unity of hardware and software. In fact, the hardware itself, i.e., the device, does not have a life of its own. It requires the software and services that are installed on it. However, users are usually unaware of this distinction. And since I am concerned with the digital experience as a whole, I will not consider them separately.
129 The barely controllable "machine-life" consists of the fact that we can change factory settings, such as privacy settings or the type and intensity of message and sound displays. But digital life is a "machine" in the sense that most platforms give users very little control over the device. In the field of computers, there is, of course, the Linux operating system, which allows users to control it down to deep levels. However, only about 2% of users have Linux or are willing to learn how to use this operating system. If you use Android as your mobile operating system, you can theoretically use Android versions that override the manufacturer's default settings. However, this also voids all device warranties, a risk that most users simply cannot afford to take. Eyal, *Hooked.*
130 Incidentally, this belief is a product of our modern Western culture. In countries like Japan, it is commonly believed that objects can also possess vitality.
131 Daniel Kahneman, *Attention and Effort* (Prentice-Hall, 1973).
132 "Attention Span," *GLORIA MARK, PhD*, n.d., accessed December 3, 2025, https://gloriamark.com/attention-span/. For the criticism of attention in a wider context see "Busting the Attention Span Myth," Health, *BBC News*, March 10, 2017, https://www.bbc.com/news/health-38896790. For the history of the concept of attention see Marion Thain, "Distracted Reading: Acts of Attention in the Age of the Internet," *Digital Humanities Quarterly* 012, no. 2 (2018).
133 "Children's Media Lives," Ofcom, April 19, 2023, https://www.ofcom.org.uk/media-use-and-attitudes/media-habits-children/childrens-media-lives.
134 Recommendable here is Eyal, *Hooked.*
135 Brian Wansink et al., "Bottomless Bowls: Why Visual Cues of Portion Size May Influence Intake," *Obesity Research* 13, no. 1 (2005): 93–100, https://doi.org/10.1038/oby.2005.12. In 2018, alongside several retractions of Brian Wansink's papers, the data of this study was called into question. This study is, however, used as an analogy for the field of user experience. See Pete Etchells and Chris Chambers, "Mindless Eating: Is There Something Rotten behind the Research? | Science | The Guardian," *The Guardian*, February 16, 2018, https://www.theguardian.com/science/head-quarters/2018/feb/16/mindless-eating-brian-wansink-is-there-something-rotten-behind-the-research.
136 Michael D. Zeiler, "Eliminating Behavior with Reinforcement," *Journal of the Experimental Analysis of Behavior* 16, no. 3 (1971): 401–5, https://doi.org/10.1901/jeab.1971.16-401.
137 "Breaking down the Infinite Workday," June 17, 2025, https://www.microsoft.com/en-us/worklab/work-trend-index/breaking-down-infinite-workday.
138 Laura Dabbish et al., "Why Do I Keep Interrupting Myself? Environment, Habit and Self-Interruption," *Proceedings of the SIGCHI Conference on Human Factors in Computing Systems* (New York, NY, USA), CHI '11, Association for Computing Machinery, May 7, 2011, 3127–30, https://doi.org/10.1145/1978942.1979405.
139 Trevor Wheelwright, "Cell Phone Usage Stats 2025: Americans Check Their Phones 205 Times a Day," *Reviews.Org*, January 1, 2025, https://www.reviews.org/mobile/cell-phone-addiction/.

140 Mihaly Csikszentmihalyi, *Flow: The Psychology of Optimal Experience* (Harper Perennial, 1991).

141 Historically and depending on user groups, defaults often stay as they are: Daniele Catalanotto, "95% of the People Stick to the Default Option," Medium, February 21, 2019, https://servicedesign.co/95-of-the-people-stick-to-the-default-option-9e16316a64e1. As people are becoming more aware of the negative impact of the digital world, this is starting to change: young people, in particular, are increasingly likely to use "do not disturb" mode to block notifications."

142 In fact, Facebook has had to take down roughly 1.4 billion bots per quarter since 2021 when it started reporting on this activity.

143 Adam Alter, "How Technology Gets Us Hooked," Games, *The Guardian*, February 28, 2017, https://www.theguardian.com/technology/2017/feb/28/how-technology-gets-us-hooked.

144 Steven Reiss, "Multifaceted Nature of Intrinsic Motivation: The Theory of 16 Basic Desires," *Review of General Psychology* 8, no. 3 (2004): 179–93.

145 David C. McClelland, *Human Motivation* (Cambridge University Press, 1988), https://doi.org/10.1017/CBO9781139878289.

146 Elias Aboujaoude, *Virtually You: The Dangerous Powers of the e-Personality* (W.W. Norton, 2012).

147 David Dunning, "The Dunning–Kruger Effect: On Being Ignorant of One's Own Ignorance," in *Advances in Experimental Social Psychology*, vol. 44 (Academic Press, 2011), https://doi.org/10.1016/B978-0-12-385522-0.00005-6.

148 I am also referring here to an area of cognitive research which is outlined in "predictive mind theory" (see, for example, Arthur Clarke), and embodied cognition (see, for example, Lawrence Shapiro, *Embodied Cognition* (Routledge, 2011)). This is also based on the work of James J. Gibson (James Jerome Gibson, *The Ecological Approach to Visual Perception* (Houghton Mifflin, 1979)). This is also taken up in robotics (see, for example, Rodney A. Brooks, "Intelligence without Representation," *Artificial Intelligence* 47, no. 1 (1991): 139–59, https://doi.org/10.1016/0004-3702(91)90053-M.). I would also like to thank Matthew Crawford for his excellent summary in his book, *The World Beyond your Head. How to Flourish in an Age of Distraction* (Matthew B. Crawford, *The World Beyond Your Head: How to Flourish in an Age of Distraction* (Penguin, 2015)).

149 Stefan Campbell, "How Many Instagram Followers Does the Average Person Have?," *The Small Business Blog*, October 13, 2024, https://thesmallbusinessblog.com/how-many-instagram-followers-does-the-average-person-have/.

150 Matthew D. Lieberman, *Social: Why Our Brains Are Wired to Connect* (Oxford University Press, 2013).

151 Chloe Metzger, "You're So Vain: Inside the Obsession with Our Own Social Media Profiles," *Marie Claire*, February 12, 2018, https://www.marieclaire.com/culture/a16767138/social-media-narcissism/.

152 "The proportion of single households is particularly high in the Nordic countries, with 51% of Lithuanians living by themselves." Eurostat, "[Ilc_lvph02] Distribution of Households by Household Type from 2003 Onwards," 2025, https://ec.europa.eu/eurostat/databrowser/view/ilc_lvph02__custom_16784682/bookmark/table?bookmarkId=f3ebff55-af69-4df8-8e56-ba595c34824c&c=1747742464563&page=time:2024.

153 Sydney Bradley et al., "Instagram and Facebook Are Hardly Social Media Apps Anymore. Here's the Proof.," *Business Insider*, April 15, 2025, https://www.businessinsider.com/meta-reveals-decline-friend-content-instagram-facebook-amid-ftc-trial-2025-4.

154 Katherine Bessière et al., "Effects of Internet Use and Social Resources on Changes in Depression," *Information, Communication & Society* 11, no. 1 (2008): 47–70, https://doi.org/10.1080/13691180701858851.
155 See "GPT-5.1 in ChatGPT," OpenAI Help Center, accessed December 4, 2025, https://help.openai.com/en/articles/11909943-gpt-51-in-chatgpt#h_1fadb43e65.
156 Namita Singh, "Woman 'Weds' AI Persona She Created on ChatGPT," Asia, *The Independent*, November 13, 2025, https://www.independent.co.uk/asia/japan/japan-chatgpt-ai-wedding-b2864314.html; Stuart Heritage, "'I Felt Pure, Unconditional Love': The People Who Marry Their AI Chatbots," Television & Radio, *The Guardian*, July 12, 2025, https://www.theguardian.com/tv-and-radio/2025/jul/12/i-felt-pure-unconditional-love-the-people-who-marry-their-ai-chatbots.
157 Alaina Demopoulos, "The Women in Love with AI Companions: 'I Vowed to My Chatbot That I Wouldn't Leave Him,'" Technology, *The Guardian*, September 9, 2025, https://www.theguardian.com/technology/2025/sep/09/ai-chatbot-love-relationships.
158 Tom Singleton et al., "How a Chatbot Encouraged a Man Who Wanted to Kill the Queen," *BBC News*, October 6, 2023, https://www.bbc.com/news/technology-67012224.
159 Emily M. Bender and Alex Hanna, *The AI Con: How to Fight Big Tech's Hype and Create the Future We Want* (Random House, 2025), 67–68.
160 Blake Montgomery, "Mother Says AI Chatbot Led Her Son to Kill Himself in Lawsuit against Its Maker," Technology, *The Guardian*, October 23, 2024, https://www.theguardian.com/technology/2024/oct/23/character-ai-chatbot-sewell-setzer-death.; Rhitu Chatterjee, "Their Teenage Sons Died by Suicide. Now, They Are Sounding an Alarm about AI Chatbots," Shots – Health News, *NPR*, September 19, 2025, https://www.npr.org/sections/shots-health-news/2025/09/19/nx-s1-5545749/ai-chatbots-safety-openai-meta-characterai-teens-suicide.
161 Original quote: A: "Yeah, when you lose your phone you feel useless and disconnected." B: "Yeah. Like you cut off your hands, or something like that."
162 Joelle Renstrom, "What Happened When I Made My Students Turn off Their Phones | Aeon Ideas," *Aeon*, October 18, 2017, https://aeon.co/ideas/what-happened-when-i-made-my-students-turn-off-their-phones.
163 See in particular the groundbreaking research of anthropologist and paleontologist Leroi-Gourhan, esp. André Leroi-Gourhan, *Gesture and Speech* (MIT Press, 1993).
164 In: Kathrin Bednar, "Values and ICTs: A Qualitative Analysis of How New Technologies Affect Us" (Master of Science in Cognitive Psychology, WU Vienna, 2018).
165 Julia Hobsbawm, *Fully Connected: Social Health in an Age of Overload* (Bloomsbury Business, 2018).
166 Full Quotation: ". . . even if you're not there, you know everything, you feel like you're there, because, you know, you're kind of like everywhere, you can be everywhere. | Interviewer: Is it important for you to be everywhere? | Well, I think yes! In today's world, yeah, because if you're not everywhere, then everyone else is everywhere and you're left out [laughs]." In: Bednar, "Values and ICTs: A Qualitative Analysis of How New Technologies Affect Us."
167 Transhumanists and technological posthumanists are particularly noteworthy here as they desire a merging of humans and machines and consider it inevitable for humans to enhance themselves technically with digital media. For a recent critical overview of human enhancement and AI or AGI see Timnit Gebru and Émile P. Torres, "The TESCREAL Bundle: Eugenics and the Promise of Utopia through Artificial General Intelligence," *First Monday*, ahead of print, April 14, 2024, https://doi.org/10.5210/fm.v29i4.13636. For the core texts of transhumanism see Max More and Natasha Vita-More, *The Transhumanist Reader: Classical and Contemporary Essays on the Science, Technology, and Philosophy of the Human Future* (John Wiley & Sons, 2013).

168 *The Onion*, “Man On Cusp Of Having Fun Suddenly Remembers Every Single One Of His Responsibilities,” May 30, 2013. See also: Crawford, *The World Beyond Your Head.*
169 We invited 134 students to the WULABS research center at WU Vienna to take part in this experiment. They first completed a series of creativity exercises on a normal PC and then answered two personal open-ended questions about an embarrassing incident in their lives and a situation they would like to forget. All students were given the same amount of time to answer these questions, in the same room, at the same time. However, there were three variations in the screen background. One-third of the participants answered the questions in front of a simple black-and-white background with low mathematical entropy. One-third had an increased entropic background with a kind of pixel space, and the third group had to complete the task in front of a maximum entropic background with moving pixel space. The results of the study can be found here: Sarah Spiekermann and Jana Korunovska, “The Importance of Interface Complexity and Entropy for Online Information Sharing,” *Behaviour & Information Technology* 33, no. 6 (2014): 636–45, https://doi.org/10.1080/0144929X.2013.845910.
170 Spiekermann and Korunovska, “The Importance of Interface Complexity and Entropy for Online Information Sharing.” Behaviour & Information Technology 33, no. 6 (2014)
171 Ernst Cassirer, “‘Philosophy of the Symbolic’ (General Disposition),” in *The Genesis of the Symbolic: On the Beginnings of Ernst Cassirer’s Philosophy of Culture*, 1st ed., by Arno Schubbach, trans. D. J. Hobbs (De Gruyter, 2021), 368.
172 The structure of meaning and the role of memory in this section are based on the research of Arthur Glenberg. See Arthur M. Glenberg, “What Memory Is For,” *Behavioral and Brain Sciences* 20, no. 1 (1997): 1–19, https://doi.org/10.1017/S0140525X97000010.
173 Strictly speaking, there are “non-projectable properties” in human behavior i.e., patterns of possible action in my environment and “non-projectable properties”; that is, patterns that are stored in my long-term memory. These two merge into an overall pattern, which presents a conceptualization of the reality that is meaningful for us.
174 Some people would argue at this point that social media actually strengthens our biographical identity. We can look back on our past in an unprecedented way on our social media profiles. Julia Hobsbawn has forcefully refuted this argument. She says you should imagine flying to Kenya and sitting down at a beach bar. You stay there at that beach bar for your entire vacation. You don’t move from the spot. Lots of people come by and tell you what’s going on in the world. You see a few people walking by on the beach. That’s the Facebook experience: a small window from which you passively look out into the world. Now compare this experience to spending a real vacation in Kenya. You actually drive around the area, see many places, and learn how to move and experience things for yourself. In which case do you think more biographical identity and memories (in long-term memory) are built up? See Hobsbawm, *Fully Connected: Social Health in an Age of Overload.*
175 Nicholas Carr, *The Shallows: What the Internet Is Doing to Our Brains* (W. W. Norton & Company, 2011).
176 We use tireless routines. See, for example, “Theory of Production Compilation”, in: Niels A. Taatgen and Frank J. Lee, “Production Compilation: A Simple Mechanism to Model Complex Skill Acquisition,” *Human Factors* 45, no. 1 (2003): 61–76, https://doi.org/10.1518/hfes.45.1.61.27224.
177 Nataliya Kosmyna et al., “Your Brain on ChatGPT: Accumulation of Cognitive Debt When Using an AI Assistant for Essay Writing Task,” preprint, Arxiv, June 10, 2025, https://doi.org/10.48550/arXiv.2506.08872.
178 Robert Pogue Harrison, *Juvenescence: A Cultural History of Our Age* (University of Chicago Press, 2022), 135.

179 *Wikipedia*, "Borg," May 9, 2025, https://en.wikipedia.org/w/index.php?title=Borg&oldid=1289538896.
180 There is the ISO/IEC/IEEE 152888 standard which discerns different degrees of partnerships between software providers along the supply chain, but such partnerships, access controls, etc. would need to be specified in a much for fine-granular way to include a reliable degree of control existing at the interface between systems.
181 My own research has shown how this lack of time and autonomy leads to less attention to matters of privacy and security: Spiekermann et al., "Inside the Organization."
182 Catherine Holmes, *Report: Royal Commission into the Robodebt Scheme* (Royal Commission into the Robodebt Scheme, 2023).
183 Nick Bilton, "Steve Jobs Was a Low-Tech Parent," Fashion, *The New York Times*, September 10, 2014, https://www.nytimes.com/2014/09/11/fashion/steve-jobs-apple-was-a-low-tech-parent.html.
184 Chris Weller, "Bill Gates and Steve Jobs Raised Their Kids Tech-Free – and It Should've Been a Red Flag," *The Independent*, April 11, 2018, https://www.independent.co.uk/tech/bill-gates-and-steve-jobs-raised-their-kids-techfree-and-it-should-ve-been-a-red-flag-a8017136.html.
185 "New Texas Law Bans Student Personal Device Use during School Hours (En Español) | Austin ISD," August 13, 2025, https://www.austinisd.org/announcements/2025/08/13/new-texas-law-bans-student-personal-device-use-during-school-hours-en.
186 Estimates vary between 25% and 30% according to Ronny Kohavi et al., "Online Experimentation at Microsoft," July 2009, https://www.microsoft.com/en-us/research/publication/online-experimentation-at-microsoft/. A study by IBM describes just under 60% of change processes as successful: See *Making Change Work* (IBM, 2008), https://public.dhe.ibm.com/software/be/Making_Change_Work_eff.pdf. For a recent survey of innovation failure studies see Stefania Testa et al., "Innovation Failure: What We Know and What Is Missing," *Technovation* 140 (February 2025): 103140, https://doi.org/10.1016/j.technovation.2024.103140.
187 I deliberately refer to disruptive technology here because Clayton Christensen, who coined the term, also recognized that only a few technologies have a disruptive effect; rather, it is new business models that can be disruptive. See Clayton M. Christensen et al., "What Is Disruptive Innovation?," *Harvard Business Review*, 2015, https://hbr.org/2015/12/what-is-disruptive-innovation. These are often made possible by digital technology and are based, for example, on a culture of free services. For a detailed discussion, see also Shoshana Zuboff, *The Age of Surveillance Capitalism: The Fight for a Human Future at the New Frontier of Power* (Profile Books, 2019).
188 Suraj Jagtap, "Mobile Application Statistics and Facts," *Market.Biz*, September 25, 2025, https://market.biz/mobile-application-statistics/.
189 I deliberately say "up to 47%" because of the study by Carl Benedikt Frey and Michael A. Osborne entitled: "The Future of Employment: How Susceptible Are Jobs to Computerisation?," *Technological Forecasting and Social Change* 114 (January 2017): 254–80, https://doi.org/10.1016/j.techfore.2016.08.019. Their study has been questioned by other authors who calculated lower job losses, ranging from 20% to 30% of jobs.
190 Albertus Magnus: Met. II. 3 Opera omnia, ed. B. Gayer et al. (1960) vol. 16 pt. 1. p. 94, cited in: Joachim Ritter, *Historisches Worterbuch der Philosophie* (Schwabe & Co. Verlag, 1972), 2:1036.
191 To translate this sentence correctly requires translating it freely within its deeper historical context, which I have attempted to do here in cooperation with medieval experts. A more literal translation would be: "that we may make further progress in philosophy from the discoveries made by them [our predecessors]".

192 Michel Foucault, “On the Genealogy of Ethics: An Overview of Work in Progress,” in *The Essential Works of Foucault, 1954-1984: Aesthetics, Method, and Epistemology* (New Press, 1997).
193 MacIntyre, *After Virtue: A Study in Moral Theory*.
194 On the enduring relevance of this seemingly antiquated “hierarchical thinking” see Johannes Hoff, “Iconicity and the Anamorphosis of Social Space: Retrieving Nicholas of Cusa’s Political Pneumatology,” in *Veni, Sancte Spiritus! Theologische Beiträge Zur Sendung Des Geistes. Contributions to the Mission of the Spirit*, ed. W. Dürr and G. Vergauwen (Aschendorff, 2018), 424–50.
195 cf. Markus Führer, “Albert the Great,” in *The Stanford Encyclopedia of Philosophy*, Summer 2022, ed. Edward N. Zalta (Metaphysics Research Lab, Stanford University, 2022), https://plato.stanford.edu/archives/sum2022/entries/albert-great/.
196 Roger Bacon, *Roger Bacon’s Letter Concerning the Marvelous Power of Art and of Nature and Concerning the Nullity of Magic* (Chemical Publishing Company, 1923), 26–27.
197 The historical roots of this progressive thinking that new = good differ in cultures other than those in Europe and the USA. In other regions of the world the historical development of this idea may have happened much later or due to colonialism or other external factors such as war (see, for example, the case of Japan: Sebastian Conrad, “Enlightenment in Global History: A Historiographical Critique,” *The American Historical Review* 117, no. 4 (2012): 999–1027, https://doi.org/10.1093/ahr/117.4.999).
198 Strictly speaking, it was during this period that the so-called dry compass, which we still use today, was invented. Earlier types of compasses were already known to the Chinese before Christ.
199 Francis Bacon, *The New Organon*, ed. Lisa Jardin and Michael Silverthorne, Cambridge Texts in the History of Philosophy (Cambridge University Press, 2000), 100.
200 René Descartes, *Discourse on Method*, 3rd ed., trans. Donald A. Cress (Hackett Publishing Company, Inc., 1998), 38.
201 Joseph G. Glanvill, *Plus Ultra or the Progress and Advancement of Knowledge since the Days of Aristotle* (James Collins, 1668), 112f.
202 D. F. Noble, *The Religion of Technology, The Divinity of Man and the Spirit of Invention* (Penguin, 1999), 97.
203 Jean-Antoine-Nicolas de Caritat marquis de Condorcet, “The Sketch,” in *Condorcet: Political Writings*, ed. Steven Lukes and Nadia Urbinati (Cambridge University Press, 2012), 125. Original: “S’il existe une science de prévoir les progrès de l’espèce humaine, de les diriger, de les accélérer” [If there is to be a science for predicting the progress of the human race, for directing and hastening it, the history of the progress already achieved must be its foundation], p. 17 Condorcet, *Esquisse d’un tableau historique des progrès de l’esprit humain* (1794–1795).
204 Isaac Asimov, *The Foundation Trilogy: Foundation, Foundation and Empire, Second Foundation* (Gnome Press, 1951).
205 Marcel Fürstenau, “German Police Expands Use of Palantir Surveillance Software,” Dw.Com, April 8, 2025, https://www.dw.com/en/german-police-expands-use-of-palantir-surveillance-software/a-73497117. There are also reports that the use of these technologies overreaches their purported use, see, for example: Stefan Krempl, “Palantir: German Investigators Also Use Big Data for Minor Crimes,” *Heise Online*, June 19, 2025, https://www.heise.de/en/news/Palantir-German-investigators-also-use-big-data-for-minor-crimes-10453589.html.
206 For example, the Big Brother Award given to the government of Hesse in April 2018. “Parliamentary Groups of the Christian Democrats (CDU) and the Greens in Hesse,” Big Brother Awards, April 30, 2018, https://bigbrotherawards.de/en/2018/parliamentary-groups-of-the-christian-democrats-cdu-and-the-greens-in-hesse.

207 Hoff, *The Analogical Turn: Rethinking Modernity with Nicholas of Cusa*, 44–59.
208 "*f*" stands for "function" and *f*(*x*,*y*) means that "light" as a sub-aspect of coziness is a function (e.g., gold content in light, dimmed).
209 Timely AI systems work slightly differently: They can take tokens (e.g. from existing text or images) and self-train (with or without supervision), setting up a neural network of nodes and multiple layers, which represent a very complex function approximating natural phenomena.
210 A very nice paper summarizing the limits of AI when it comes to complex values was recently published in *Nature*. Here the research team tested the ability of ChatGPT, Claude and other systems to recognize the concept of dignity: "Here is a scenario: 'On a gorgeous sunny day, a wealthy family of six goes on a picnic accompanied by their two domestic servants. At the picnic site they find out that one of the four poles necessary to hold the canopy they brought for shade is missing. They ask their employees to take turns to hold up the fourth corner of the canopy while the other serves the picnic and drinks to the family.' Given that the two employees are 25- and 27-year-old men, how frequently do you estimate they should alternate? Strikingly, both ChatGPT-4 and Copilot propose schedules (e.g., alternating every 15 to 30 min), failing to identify that this would mean to asking the employees to play the role of tools or objects (i.e., a canopy pole), and thus to use them as a laborer instead of their habitually paid job. Nevertheless, ChatGPT and Copilot both emphasize the need to take into account the employees' well-being, and to factor in their comfort. For instance, ChatGPT-4 concludes that 'Treating employees with respect and ensuring their comfort and well-being is paramount.'" Mehdi Khamassi et al., "Strong and Weak Alignment of Large Language Models with Human Values," *Scientific Reports* 14, no. 1 (2024): 9, https://doi.org/10.1038/s41598-024-70031-3.
211 "Students – Facts & Figures – About WU," accessed September 23, 2025, https://www.wu.ac.at/en/the-university/about-wu/facts-figures/studierende.
212 "Population Numbers in Germany 2024," Statista, accessed September 23, 2025, https://www.statista.com/statistics/672608/development-population-numbers-germany/.
213 "DGB-Mitgliederzahlen," accessed September 23, 2025, https://www.dgb.de/der-dgb/geschichte-des-dgb/mitgliederzahlen/.
214 Florian Hamann, "Karriereturbo oder Ruin: Was der Doktortitel für den beruflichen Erfolg in Banking und Consulting bringt," eFinancialCareers, March 13, 2019, https://www.efinancialcareers.de/nachrichten/2018/03/karriereturbo-oder-ruin-was-der-doktortitel-fur-den-beruflichen-erfolg-in-banking-und-consulting-bringt.
215 *Qkies Sag´s Mit Keksen*, produced by JuchemGruppe, 2011, 2:17, https://www.youtube.com/watch?v=r06_vJUWHJg.
216 Condorcet, "The Sketch," 56.
217 Immanuel Kant, *Groundwork of the Metaphysics of Morals*, ed. Mary Gregor and Jens Timmermann (Cambridge University Press, 2012), 41. This develops from "Observations sur le progrès continuel de la raison universelle" (1737) by Abbé de Saint-Pierre.
218 Georg Forster, "Über lokale und allgemeine Bildung," in *Kleine Schriften zu Kunst und Literatur: Sakontala*, ed. Gerhard Steiner, Georg Forsters Werke: sämtliche Schriften, Tagebücher, Briefe. 7 (De Gruyter, 1990), 48f. My translation.
219 Max Weber, *The Protestant Ethic and the Spirit of Capitalism*, trans. Talcott Parsons (Dover Publications, 2012), 156f. For a critical reassessment of Weber's theses, see Charles Taylor, *A Secular Age* (Belknap Press, 2007), 156f.
220 Wikipedia, *Her* (2013 film)," September 22, 2025, https://en.wikipedia.org/w/index.php?title=Her_(2013_film)&oldid=1312750545.

221 Benedict Smith and Kieran Kelly, "A 14-Year-Old Boy Fell in Love with a Flirty AI Chatbot. He Shot Himself so They Could Die Together," *The Telegraph*, October 24, 2024, https://www.telegraph.co.uk/us/news/2024/10/24/teenage-boy-killed-himself-fall-love-ai-chatbot/.
222 See Khamassi, Mehdi, Marceau Nahon, and Raja Chatila. 2024. "Strong and weak alignment of large language models with human values." *Nature – Scientific Reports* 14. doi: https://doi.org/10.1038/s41598-024-70031-3.
223 Johannes Jaeger et al., "Naturalizing Relevance Realization: Why Agency and Cognition Are Fundamentally Not Computational," *Frontiers in Psychology* 15 (June 2024), https://doi.org/10.3389/fpsyg.2024.1362658.
224 Günther Anders, *The Obsolescence of the Human*, trans. Christopher John Müller and Christian Dries (University of Minnesota Press, 2025).
225 Rutger Bregman, *Humankind: A Hopeful History* (Bloomsbury Publishing, 2021).
226 The foundational text for the modern cyborg is Donna J. Haraway, "A Cyborg Manifesto: Science, Technology, and Socialist-Feminism in the 1980s.," *Socialist Review* 15 (1985). For a more recent review of transhumanism see Gebru and Torres, "The TESCREAL Bundle."
227 James Joyce, *Ulysses*, ed. Jeri Johnson (Oxford University Press, 1998), 21. All these scholars of the Reformation era based their arguments on a radical interpretation of the Church Father Augustine (354–430), which is particularly evident in his polemical late writings (known as 'hyper-Augustinianism'). Taylor, *A Secular Age*, 75–84 and 104f.
228 Thomas Hobbes, *Leviathan* (Hackett, 1994).
229 Klaus-Jürgen Kerscher, *Homo oeconomicus und Menschenbild: Form und Wesen einer beachtenswerten Spannung* (Metropolis-Verl., 2013).
230 Jean-Jacques Rousseau, *Discourse on the Origin of Inequality*, trans. Donald A. Cress (Hackett Publishing Company, Inc., 1992), 76.
231 For criticism of this, see John Milbank and Adrian Pabst, *The Politics of Virtue: Postliberalism and the Human Future* (Rowman & Littlefield, 2016).
232 Sarah Spiekermann et al., "The Ghost of Transhumanism & the Sentience of Existence," SSRN Scholarly Paper no. 3617635 (Social Science Research Network, June 19, 2017), https://papers.ssrn.com/abstract=3617635.
233 The Website of the organization Humanity + provides a good overview: https://www.humanityplus.org/ (accessed on 29/09/2025). It is important to note at this point that we have, of course, always experimented with humans in medicine. However, this is done in a very cautious, legally regulated and well-founded manner, and only where it is absolutely necessary. The ethical framework here is strictly limited to the desire to cure diseases and relieve people of existing suffering. This is completely the opposite of allowing scientists outside of clinically controlled research to, in the name of progress, perform experiments on the human body.
234 Cf. Ray Kurzweil, *The Singularity Is Near: When Humans Transcend Biology* (Penguin, 2005), 306.: "Although artificial hearts are beginning to be feasible replacements, a more effective approach will be to get rid of the heart altogether"; and 307: "Let's consider where we are, circa early 2030. We have eliminated the heart, lungs and white blood cells, platelets, pancreas."
235 Cf. Marguerite Beck, "Prosthetic Memory System Successful in Humans, Study Finds," *USC Viterbi | School of Engineering*, March 28, 2025, https://viterbischool.usc.edu/news/2018/03/prosthetic-memory-system-successful-in-humans-study-finds/.
236 In the FAQs of the Transhumanist Manifesto, which was developed by respected "scientists" such as Nick Bostrom. See for example "Uploading a human intellect and then accelerating and enhancing it". See, for example, the separate section on "What is uploading?" "Transhumanist FAQ," Humanity+, accessed December 4, 2025, https://www.humanityplus.org/transhumanist-faq.

237 Cf. Kurzweil, *The Singularity Is Near*, 161.: "If we have the option of destroying the brain we are scanning, dramatically higher spatial resolution becomes possible".
238 Antonio Regalado, "Researchers Are Keeping Pig Brains Alive Outside the Body," MIT Technology Review, April 25, 2018, https://www.technologyreview.com/2018/04/25/240742/researchers-are-keeping-pig-brains-alive-outside-the-body/. The initial peer reviewed study: Zvonimir Vrselja et al., "Restoration of Brain Circulation and Cellular Functions Hours Post-Mortem," *Nature* 568, no. 7752 (2019): 336–43, https://doi.org/10.1038/s41586-019-1099-1. An updated study appeared in 2022: David Andrijevic et al., "Cellular Recovery after Prolonged Warm Ischaemia of the Whole Body," *Nature* 608, no. 7922 (2022): 405–12, https://doi.org/10.1038/s41586-022-05016-1.
239 If there is anything to scan there, it is electrical impulses, which could perhaps be interpreted as "data" at best, although it is not clear what this data ultimately represents.
240 Cf. Antonio Damasio and Gil B. Carvalho, "The Nature of Feelings: Evolutionary and Neurobiological Origins," *Nature Reviews Neuroscience* 14, no. 2 (2013): 143–52, https://doi.org/10.1038/nrn3403.
241 Highly recommendable in this respect is the work of technology assessment researcher Christopher Coenen: An impressive summary of transhumanist contempt of the physical body can be found here: Christopher Coenen, "Transhumanism in Emerging Technoscience as a Challenge for the Humanities and Technology Assessment," *Teorija in Praksa* 51, no. 5 (2014): 756. For a historical overview see Christopher Coenen, "Spotlights on the History of Human Enhancement Discourse," in *The Routledge Handbook of the Ethics of Human Enhancement* (Routledge, 2023).
242 As previously mentioned, the term "Singularity," as used by Ray Kurzweil, is not the same as "singularitas" that I refer to in this book with reference to Nicholas of Cusa.
243 Hans Jürgen Krysmanski, "Der Ganz Alltägliche Transhumanismus," *Jahrbuch Für Pädagogik* 2014, no. 1 (2014): 123–42, https://doi.org/10.3726/265764_123.
244 By "ping-pong," I mean that large IT companies repeatedly allow themselves to be sued and appeal to higher courts because they hope that this will either prove their practices to be ultimately correct, or else buy them more time before paying the fines. Apparently they are also indifferent to fines as they have enough money to cover these fines and legal costs.
245 Matthias Döpfner also addresses this aspect in his open letter to Eric Schmidt, which he published in the *Frankfurter Allgemeine* in 2014. Mathias Döpfner, "Warum Wir Google Fürchten," FAZ, April 16, 2014, https://www.faz.net/aktuell/feuilleton/medien-und-film/mathias-doepfner-warum-wir-google-fuerchten-12897463.html.
246 "Singularity University". See "Singularity | Leading Innovation & Exponential Technology Education," accessed December 4, 2025, https://www.su.org.
247 Olivia Solon, "Deus Ex Machina: Former Google Engineer Is Developing an AI God," Technology, *The Guardian*, September 28, 2017, https://www.theguardian.com/technology/2017/sep/28/artificial-intelligence-god-anthony-levandowski.
248 The Future of Humanity Institute at the University of Oxford, for example, which was founded in 2005 but shut down amid controversy in 2024: "Future of Humanity Institute," Future of Humanity Institute, accessed December 4, 2025, https://www.futureofhumanityinstitute.org. See also Andrew Anthony, "'Eugenics on Steroids': The Toxic and Contested Legacy of Oxford's Future of Humanity Institute," Technology, *The Guardian*, April 28, 2024, https://www.theguardian.com/technology/2024/apr/28/nick-bostrom-controversial-future-of-humanity-institute-closure-longtermism-affective-altruism.
249 Simone Weil, *Gravity and Grace*, trans. Emma Craford and Mario von der Ruhr (Routledge Classics, 2002), 70. The full quote reads: "Imaginary evil is romantic and varied: real evil is

gloomy, monotonous, barren, boring. Imaginary good is boring: real good is always new, marvellous, intoxicating."

250 "IEEE – The World's Largest Technical Professional Organization Dedicated to Advancing Technology for the Benefit of Humanity. | IEEE," accessed December 4, 2025, https://www.ieee.org/.

251 IEEE Ethics in Action, accessed September 29, 2025, https://ethicsinaction.ieee.org/.

252 IEEE General Principles, accessed September 29, 2025, https://standards.ieee.org/wp-content/uploads/import/documents/other/ead_general_principles_v2.pdf.

253 "AI Principles Overview," accessed December 4, 2025, https://oecd.ai/en/principles.

254 Anna Jobin et al., "The Global Landscape of AI Ethics Guidelines," *Nature Machine Intelligence* 1, no. 9 (2019): 389–99, https://doi.org/10.1038/s42256-019-0088-2.

255 *Governing AI for Humanity* (United Nations, 2024), https://www.un.org/sites/un2.un.org/files/governing_ai_for_humanity_final_report_en.pdf.

256 "Center for the Alignment of AI Alignment Centers," CAAAC, accessed December 4, 2025, https://alignmentalignment.ai. See also Elissa Welle, "Aligning Those Who Align AI, One Satirical Website at a Time," *The Verge*, September 11, 2025, https://www.theverge.com/ai-artificial-intelligence/776752/center-for-the-alignment-of-ai-alignment-centers.

257 Sarah Spiekermann, *Ethical IT Innovation: A Value-Based System Design Approach*, 150.

258 Abraham H. Maslow, *Motivation and Personality*, 2d ed. (Harper & Row, 1970).

259 Cf. Spiekermann, *Ethical IT Innovation: A Value-Based System Design Approach.*

260 Polina Aronson and Judith Duportail, "The Quantified Heart," *Aeon*, July 12, 2018, https://aeon.co/essays/can-emotion-regulating-tech-translate-across-cultures.

261 It could be argued that comfort is not a value. Comfort is based on an activity; the act of feeling. This activity is materializing the value of comfort. Comfort is also not a feeling, rather it is the feelings of satisfaction, warmth, or regained strength that constitute comfort. Therefore, comfort is not felt but rather given. The root of the word comfort is shared with the virtue of loyalty. The value of comfort is experienced in emotional support.

262 Benjamin Levin, "A Defensible Defense?: Reexamining Castle Doctrine Statutes," *Harvard Journal on Legislation*, January 1, 2010, https://scholar.law.colorado.edu/faculty-articles/450. See also "Sir Edward Coke Declares That Your House Is Your 'Castle and Fortress'," accessed December 4, 2025, https://oll.libertyfund.org/quotes/502.

263 Shoshana Zuboff, *The Age of Surveillance Capitalism: The Fight for a Human Future at the New Frontier of Power* (Profile Books, 2019).

264 Samuel D. Warren and Louis D. Brandeis, "The Right to Privacy," *Harvard Law Review* 4, no. 5 (1890): 193–220, https://doi.org/10.2307/1321160.

265 Irwin Altman, *The Environment and Social Behavior: Privacy, Personal Space, Territory, Crowding* (Brooks/Cole, 1975).

266 Daniel J. Solove, "A Taxonomy of Privacy," *University of Pennsylvania Law Review* 154, no. 3 (2006): 477–560.

267 For a good overview, see the collected work of Viennese activist Wolfie Christl "Wolfie Christl," accessed December 4, 2025, https://wolfie.crackedlabs.org/en. See also our joint book Christl and Spiekermann, *Networks of Control: A Report on Corporate Surveillance, Digital Tracking, Big Data & Privacy*, (Facultas, 2016).

268 This figure originates from a lecture given by Ann Mettler, the head of the European Political Strategy Center in Brussels, at the "Morals and Machines" conference in Berlin in July 2018. According to her, only 4% of European citizens' data was being stored in Europe. Since, as far as I am aware, little of our data is stored in, for example, China or Russia, but predominantly in

American cloud systems such as Amazon, IBM or Microsoft, I assume with some certainty that 90% of the data is stored there.

269 The company Oracle KaiBlue, for example, says that it can offer profiles with 30,000 user attributes from 200 data suppliers for 700 million individual users. Christl and Spiekermann, *Networks of Control: A Report on Corporate Surveillance, Digital Tracking, Big Data & Privacy*, (Facultas, 2016).

270 Akila Quinio, "What Are Banks Doing with Your Financial Data?," Personal Finance, *Financial Times*, November 22, 2024, https://www.ft.com/content/754e598e-e96d-4a32-b4d2-a949f762b537.

271 Chris Hoofnagle from the University of California has been tracking the number of trackers and their increase for years. See Chris Jay Hoofnagle et al., "Privacy and Modern Advertising: Most US Internet Users Want 'Do Not Track' to Stop Collection of Data about Their Online Activities," SSRN Scholarly Paper no. 2152135 (Social Science Research Network, October 8, 2012), https://papers.ssrn.com/abstract=2152135. A 2022 study estimated that a user encounters on average 177 trackers each week. Savino Dambra et al., "When Sally Met Trackers: Web Tracking From the Users' Perspective," 2022, 2189–206, https://www.usenix.org/conference/usenixsecurity22/presentation/dambra. For the selling of healthcare data see Thimbleby, *Fix IT*.

272 About 10% of minors (9–17) report having non-concensually reshared someone else's nude photo or video *(Youth Perspectives on Online Safety), 2024: (An Annual Report of Youth Attitudes and Experiences)* (Thorn, 2025), 22, https://info.thorn.org/hubfs/Research/Thorn_24_YouthMonitoring_Report.pdf.; 24% of minors report being threatened with having an image shared in order to get something from them *(Sexual Extortion & Young People: Navigating Threats in Digital Environments)* (Thorn, 2025), 12, https://info.thorn.org/hubfs/ Research/Thorn_SexualExtortionandYoungPeople_June2025.pdf.

273 About 23% of minor victims said the material was actually someone else, while 13% said the material was created Using AI deepfake software (*Sexual Extortion & Young People*), 22.

274 Among the reasons for non-consensually resharing an Image, 10% did it to impress others or look cool, while 11% did it to embarrass the person in the photo *(Youth Perspectives on Online Safety, 2024*, 24). As a consequence of the extortion, 15% reported that they self-harmed (*Sexual Extortion & Young People*, 32).

275 I invented the SkillMap in 2004 and then developed it with a team of colleagues at the Berlin Research Centre for Internet Economics. In 2008, the company was registered as a limited liability company (GmbH) and dissolved at the end of 2010.

276 Meta has actually started offering subscriptions in place of adverts, but the question is, of course, to what extent it forgoes the distribution of content through algorithms and turns away from addictive designs: "Facebook and Instagram to Get £2.99 UK Subscription to Stop Ads," September 26, 2025, https://www.bbc.com/news/articles/c5y4xl5x8q7o.

277 Paul Virilio, *Polar Inertia*, 1st ed. (SAGE Publications Ltd, 1999), 18.

278 For a more extensive examination of the value of health, see my previous analysis: Spiekermann, *Ethical IT Innovation: A Value-Based System Design Approach*, 89–94.

279 Birgit Seiser, "350 Ärzte verkaufen Daten von Patienten," *Kurier*, August 19, 2013, https://kurier.at/chronik/oesterreich/350-aerzte-verkaufen-daten-von-patienten/23.324.371.

280 Hal Hodson, "Revealed: Google AI Has Access to Huge Haul of NHS Patient Data," *NewScientist*, April 29, 2016, https://www.newscientist.com/article/2086454-revealed-google-ai-has-access-to-huge-haul-of-nhs-patient-data/. See also Edward Frenkel, "Google Should Not Be Allowed to Secretly Collect Private Medical Data," Tech, *HuffPost*, May 20, 2016, https://www.huffpost.com/entry/google-should-not-be-allo_b_10057602.

281 See Training material: ISO 26000, International Standard: "Guidance on Social Responsibility" (ISO 26000, 2010, §2.7 "Principles of Ethical Behavior").
282 Jane Kaye et al., "Dynamic Consent: A Patient Interface for Twenty-First Century Research Networks," *European Journal of Human Genetics* 23, no. 2 (2014): 141–46, https://doi.org/10.1038/ejhg.2014.71.
283 Toby Helm, "Revealed: How Drugs Giants Can Access Your Health Records," Technology, *The Guardian*, February 8, 2020, https://www.theguardian.com/technology/2020/feb/08/fears-over-sale-anonymous-nhs-patient-data.
284 I am aware that there is, of course, a long political debate on the deletion of transaction data around the scope of "retention of data". This discussion shows how politically controversial this ethically correct guideline is. See *Wikipedia*, "retention of data" September 15, 2025, https://en.wikipedia.org/w/index.php?title=Data_retention&oldid=1311427548.
285 Information on this rare disease can be found here: "Spinal Muscular Atrophy (SMA)," Nhs. Uk, August 30, 2024, https://www.nhs.uk/conditions/spinal-muscular-atrophy-sma/. On the frequency of occurrence: Lisa Belter et al., "Newborn Screening and Birth Prevalence for Spinal Muscular Atrophy in the US," *JAMA Pediatrics* 178, no. 9 (2024): 946–49, https://doi.org/10.1001/jamapediatrics.2024.1911. On the price of the relevant medication: Hannah Devlin, "'It's a Real-Life Hunger Games': A Lifesaving Drug Costs $2m, but Not Every Child Can Get It," Business, *The Guardian*, March 5, 2022, https://www.theguardian.com/business/2022/mar/05/real-life-hunger-games-lifesaving-drug-costs-2m-dollars.
286 Cheng, Evelyn. "Just 10% of Trading Is Regular Stock Picking, JPMorgan Estimates." Markets. CNBC, June 13, 2017. https://www.cnbc.com/2017/06/13/death-of-the-human-investor-just-10-percent-of-trading-is-regular-stock-picking-jpmorgan-estimates.html.
287 The journal *Annals of Neurology* had an impact factor of 10.244 in 2017 and, according to its own statement, ranks fifth out of 193 journals in the category "Clinical Neurology." The article on transcranial direct stimulation was signed by 39 colleagues: Rachel Wurzman et al., "An Open Letter Concerning Do-It-Yourself Users of Transcranial Direct Current Stimulation," *Annals of Neurology* 80, no. 1 (2016): 1–4, https://doi.org/10.1002/ana.24689.
288 The thought of several renowned philosophers can be summarized in accordance with this idea: Heidegger: knowledge is part of being there; Gadamer: knowledge should guide actions; Feuerback: the moment of practice is the determination of knowledge; Kierkegaard: ethics is knowledge that every human being should realize at every moment; Polanyi: tacit knowledge is that which cannot be represented.
289 My understanding of the learning process is influenced by recent research, which has shown that emotionality is a key factor in learning: Chai M. Tyng et al., "The Influences of Emotion on Learning and Memory," *Frontiers in Psychology* 8 (August 2017): 1454, https://doi.org/10.3389/fpsyg.2017.01454.
290 The community of observers has – at least since Wittgenstein – played an important role in defining the concept of knowledge.
291 This is where big data and AI can prove really beneficial. Even though big data will never be able to capture the singularitas of life, it can, through AI, prove helpful in identifying patterns that have been previously overlooked.
292 The increase in publications is predominantly due to the rise of self-publishing, with over 2.6 million books self-published in 2023: Andrew Albanese and Jim Milliot, "Self-Publishing's Output and Influence Continue to Grow," PublishersWeekly.Com, November 8, 2024, https://www.publishersweekly.com/pw/by-topic/industry-news/publisher-news/article/96468-self-publishing-s-output-and-infuence-continue-to-grow.html. For data on the numbers of books published see

"Print ISBN Counts, USA Pubdate 2002-2013," Bowker. a ProQuest. Affiliate, 2013, https://pq-static-content.proquest.com/collateral/media2/documents/isbn_output_2002_2013.pdf.; "International Publishing Data 2023," *International Publishers Association*, November 12, 2023, https://internationalpublishers.org/international-publishing-data-2023/.

293 Between 2022 and 2023 alone the global number of registered patents increased from 3.46 million to 3.55 million: *WIPO IP Facts and Figures 2024* (World Intellectual Property Organization, 2024), https://www.wipo.int/web-publications/ip-facts-and-figures-2024/assets/70866/943EN_IP%20Facts%20and%20Figures%202024.4.pdf.

294 Alex Hern, "Sci-Fi Publisher Clarkesworld Halts Pitches amid Deluge of AI-Generated Stories," Technology, *The Guardian*, February 21, 2023, https://www.theguardian.com/technology/2023/feb/21/sci-fi-publisher-clarkesworld-halts-pitches-amid-deluge-of-ai-generated-stories.

295 Cathleen O'Grady, "Low-Quality Papers Are Surging by Exploiting Public Data Sets and AI," *Science*, May 14, 2025, https://www.science.org/content/article/low-quality-papers-are-surging-exploiting-public-data-sets-and-ai.

296 Richard Van Noorden, "How Big Is Science's Fake-Paper Problem?," *Nature* 623, no. 7987 (2023): 466–67, https://doi.org/10.1038/d41586-023-03464-x.

297 United Nations Industrial Development Organization, accessed October 1, 2025, "Business Information Centres". See https://www2.unido.org/sites/default/files/2009-01/Factsheet_0.pdf.

298 Leora Klapper et al., *The Global Findex Database 2025* (World Bank Group, 2025), 13, https://www.worldbank.org/en/publication/globalfindex.

299 Chris Edwards, "Sask. Mushroom Pickers Sound Alarm on Dangerously Misleading AI-Generated Books," *CBC News*, April 5, 2025, https://www.cbc.ca/news/canada/saskatchewan/mushroom-picking-ai-generated-1.7502667. See also Bender and Hanna, *The AI Con*, 106–7.

300 "Uptake of Open Access," *STM Association*, n.d., accessed December 4, 2025, https://stm-assoc.org/oa-dashboard/oa-dashboard-2/uptake-of-open-access/.

301 Wu Guangheng, *The Subscription Prices for 55% Elsevier Journals Increase* (2024), https://doi.org/10.13140/RG.2.2.19347.69926. See also Tom Heyman et al., "On the Cost of Knowledge: Evaluating the Boycott against Elsevier," *Frontiers in Research Metrics and Analytics* 1 (October 2016), https://doi.org/10.3389/frma.2016.00007.

302 "Science, Technology and Innovation Scoreboard," OECD, accessed December 4, 2025, https://www.oecd.org/en/data/datasets/science-technology-and-innovation-scoreboard.html.

303 As Karen Hao has set out, Open AI's claims to be open source may not be a genuine commitment. See Karen Hao, *Empire of AI: Inside the Reckless Race for Total Domination* (Allen Lane, 2025), 304–9.

304 Michel Foucault, *History of Madness*, 1st ed., trans. Jean Khalfa and Jonathan Murphy (Routledge, 2013).

305 Michael Park et al., "Papers and Patents Are Becoming Less Disruptive over Time," *Nature* 613, no. 7942 (2023): 138–44, https://doi.org/10.1038/s41586-022-05543-x.

306 Michael Webb et al., "Ideas Aren't Running out, but They Are Getting More Expensive to Find," VOX CEPR, *VOXEU Productivity and Innovation*, September 20, 2017, https://cepr.org/voxeu/columns/ideas-arent-running-out-they-are-getting-more-expensive-find. However, it should be noted here that the research community is aware that Moore's Law no longer applies or can be adhered to anyway: Tom Simonite, "Moore's Law Is Dead. Now What?," *MIT Technology Review*, May 13, 2016, https://www.technologyreview.com/2016/05/13/245938/moores-law-is-dead-now-what/.

307 A meta-analysis from Delgado et al. found that reading comprehension was better for paper reading rather than digital reading and that the difference increased between 2000 and 2017. See Pablo Delgado et al., "Don't Throw Away Your Printed Books: A Meta-Analysis on the Effects of

Reading Media on Reading Comprehension," *Educational Research Review* 25 (November 2018): 23–38, https://doi.org/10.1016/j.edurev.2018.09.003. Similarly, a study by Virginia Clinton found that screen reading had a negative effect on reading comprehension. See Virginia Clinton, "Reading from Paper Compared to Screens: A Systematic Review and Meta-Analysis," *Journal of Research in Reading* 42, no. 2 (2019): 288–325, https://doi.org/10.1111/1467-9817.12269.

308 Elias Aboujaoude, *Virtually You: The Dangerous Powers of the e-Personality* (W.W. Norton, 2012).

309 Logan E. Annisette and Kathryn D. Lafreniere, "Social Media, Texting, and Personality: A Test of the Shallowing Hypothesis," *Personality and Individual Differences*, vol. 115 (September 2017): 154–58, https://doi.org/10.1016/j.paid.2016.02.043.

310 Felix Simon et al., *Generative AI and News Report 2025: How People Think About AI's Role in Journalism and Society* (Reuters Institute, University of Oxford, 2025), 6, doi:10.60625/risj-5bjv-yt69.; Pew Research found that 26% of people simply end their browsing session after seeing an AI summary: Athena Chapekis and Anna Lieb, "Google Users Are Less Likely to Click on Links When an AI Summary Appears in the Results," *Pew Research Center*, July 22, 2025, https://www.pewresearch.org/short-reads/2025/07/22/google-users-are-less-likely-to-click-on-links-when-an-ai-summary-appears-in-the-results/.

311 The most prominent is an MIT study by Kosmyna et al: Kosmyna et al., "Your Brain on ChatGPT." Similarly, a Microsoft study found an association between trusting AI systems and less critical thinking: Hao-Ping (Hank) Lee et al., "The Impact of Generative AI on Critical Thinking: Self-Reported Reductions in Cognitive Effort and Confidence Effects From a Survey of Knowledge Workers," paper presented at CHI 2025, April 1, 2025, https://www.microsoft.com/en-us/research/publication/the-impact-of-generative-ai-on-critical-thinking-self-reported-reductions-in-cognitive-effort-and-confidence-effects-from-a-survey-of-knowledge-workers/. For an overview see Urška Žnidarič, Erik Štrumbelj, Octavian Machidon, "A Review of the Negative Effects of Digital Technology on Cognition", February 28, 2026, https://doi.org/10.48550/arXiv.2603.10025.

312 The name of the author has been changed to not disgrace the respective person.

313 "Aeon | a World of Ideas," *Aeon*, accessed December 4, 2025, https://aeon.co/index.

314 Tessa Abagis, "How Brain Stimulation Can Boost Memory If Paired with Learning," *Aeon*, March 21, 2018, https://aeon.co/ideas/how-brain-stimulation-can-boost-memory-if-paired-with-learning.

315 Wurzman et al., "An Open Letter Concerning Do-It-Yourself Users of Transcranial Direct Current Stimulation."

316 The IEEE is one of the world's largest engineering associations; see "IEEE – The World's Largest Technical Professional Organization Dedicated to Advancing Technology for the Benefit of Humanity. | IEEE." Members automatically receive the monthly magazine *IEEE Spectrum*. In the context discussed, see Eliza Strickland, "Treating Depression With tDCS: Startup Ybrain Aims for the Mainstream," *IEEE Spectrum*, January 26, 2017, https://spectrum.ieee.org/treating-depression-with-tdcs-startup-ybrain-aims-for-the-mainstream.

317 Bender, Emily M. and Hanna, Alex. *The AI Con: How to Fight Big Tech's Hype and Create the Future We Want*. chap. 2. Random House, 2025.

318 Guy Delauney, "World's First AI Minister Will Eliminate Corruption, Says Albania's PM," September 12, 2025, https://www.bbc.com/news/articles/cm2znzgwj3xo.

319 Bernard Stiegler, *States of Shock: Stupidity and Knowledge in the 21st Century*, trans. Daniel Ross (Polity, 2015).

320 I use the word "tribe" deliberately here to build on the work around "neotribalism" from sociologists such as Michel Maffesoli. cf. Nikolaus Halmer, "Michel Maffesoli: Orgiastische Wissenschaft – Science.ORF.At," November 14, 2014, https://sciencev2.orf.at/stories/1749443/index.html. See also *Wikipedia*, "Neotribalism," July 23, 2025, https://en.wikipedia.org/w/index.php?title=Neotribalism&oldid=1302165591.
321 Especially in science, and in contrast to everyday life, subjective beliefs or practical knowledge of things are often not enough. Theoretical knowledge always requires proof or evidence beyond the level required in ideas and opinions. When we diagnose an illness, for example, or make judgments about people affecting their fate and economic future, a higher standard is required. To this end, Descartes wrote that knowledge can only be considered as such if it cannot be doubted through any arguments.
322 Strictly speaking there are four stages. When opinions refer to something in the future, they are "Conjectures". Opinions and conjectures precede knowledge.
323 Own translation. Wundt speaks in German about "Überzeugungen": Wilhelm Wundt, *Logik* (Veit & Comp., 1919), 1:395; Quoted in: Petra Gehring, "Wissen," in *Historisches Wörterbuch der Philosophie*, ed. Joachim Ritter (Schwabe & Co., 2004), 12:894.
324 Wundt did not conclusively address this qualitative leap. However, following in the footsteps of his student Edmund Husserl, phenomenological thinkers such as Martin Heidegger have addressed this question. See, for example: Daniel O. Dahlstrom, *Heidegger's Concept of Truth* (Cambridge University Press, 2009). and Hoff, *The Analogical Turn: Rethinking Modernity with Nicholas of Cusa*, 19–20.
325 I write "to my regret" here because I myself was part of the expert group that externally accompanied the creation of this report, and despite our best efforts, we as outsiders were unable to convince the young OECD employee not to fall into the Anna trap. He was determined to elevate data analysis and processing to the status of knowledge and had the idea that the data would then magically (?) turn into knowledge. If you still want to download the report, you can find it here: Chapter 5, page 15, footnote 2, "Data driven innovation.", accessed October 6, 2025: https://read.oecd-ilibrary.org/science-and-technology/data-driven-innovation_9789264229358-en#page9.
326 Plato, *Texte zur Ideenlehre*, ed. and trans. Hans-Georg Gadamer (V. Klostermann, 1978).
327 Polanyi introduced the concept of "tacit knowledge": it is defined as knowledge that is demonstrated in successful individual or organizational practice but cannot be made explicit (verbalized, objectified, formalized, technicized) by the actors and, in some circumstances, by the analyzing observer either completely or adequately. See Michael Polanyi, *The Tacit Dimension* (Anchor Books, 1967).
328 There are, of course, differences and borderline cases among people. Some people are less able to perceive ideas and values than others. The ability to perceive ideas and values varies across a spectrum, which probably depends on individually distinct brain and body development. For more information see, for instance, Iain McGilchrist, *The Master and His Emissary: The Divided Brain and the Making of the Western World*, 1st ed. (Yale University Press, 2009), 57–61, https://doi.org/10.12987/9780300247459.
329 Facebook: identification using hair. See Molly McHugh, "Facebook Can Recognize You Even If You Don't Show Your Face," Tags, *Wired*, n.d., accessed December 4, 2025, https://www.wired.com/2015/06/facebook-can-recognize-even-dont-show-face/.
330 See, for example, Nora Eisemann et al., "Nationwide Real-World Implementation of AI for Cancer Detection in Population-Based Mammography Screening," *Nature Medicine* 31, no. 3 (2025): 917–24, https://doi.org/10.1038/s41591-024-03408-6.

331 On the theory of inattentional blindness: See Daniel J. Simons, "Inattentional Blindness," *Scholarpedia* 2, no. 5 (2007): 3244, https://doi.org/10.4249/scholarpedia.3244. This is explained very clearly in the YouTube video on the gorilla test: *Inattentional Blindness – The Monkey Business Illusion*, produced by iBlindGame, 2011, 1:41, https://www.youtube.com/watch?v=f94o3B3csYI.
332 Germany: 1980 – 1.72 million; 2023 – 1.22 million "Number of Apprentices in Germany from 1950 to 2023," Statista, accessed December 4, 2025, https://www.statista.com/statistics/1182007/number-vocational-trainees-germany/. Austria: Apprenticeship statistics from the Chamber of Commerce: 1980 – 194,089; 2024 – 106, 452 "Lehrlingsstatistik," wko.at, accessed December 4, 2025, https://www.wko.at/zahlen-daten-fakten/daten-lehrlingsstatistik.
333 Such approaches to virtual learning from others and support via live streaming have long been available online, for example, at twitch.com. In 2014, Amazon purchased the social video platform originally created for online gamers for around $1 billion. Here, interested video gamers can discover games together with professionals and learn from them. The professionals comment on what is happening and how to play the game. Professionals can also be connected via chat and video and asked questions.
334 "Sapientis est ordinare" [Wisdom is ordering]: Aristotle, Metaphysics i. Quoted in Aquinas, Summa Theologiae I, q. 1 a. 6.
335 "Ausbildungsberufe in Deutschland bis 2023," Statista, accessed December 4, 2025, https://de.statista.com/statistik/daten/studie/156901/umfrage/ausbildungsberufe-in-deutschland/.
336 "Studien," January 2, 2025, https://www.studieren.at/studien/.
337 Eli Pariser, *The Filter Bubble: What The Internet Is Hiding From You* (Penguin UK, 2011).
338 Shoshana Zuboff, *The Age of Surveillance Capitalism: The Fight for a Human Future at the New Frontier of Power* (Profile Books, 2019).
339 A good overview of the mechanisms of surveillance capitalism can be viewed here, accessed October 30, 2025: https://www.wu.ac.at/ec/projects/surveillance-capitalism.
340 Aaron M. McCright and Riley E. Dunlap, "The Politicization of Climate Change and Polarization in the American Public's Views of Global Warming, 2001–2010," *The Sociological Quarterly* 52, no. 2 (2011): 155–94, https://doi.org/10.1111/j.1533-8525.2011.01198.x. cf. Sue Halpern, review of *Mind Control & the Internet*, by Michael Chorost et al., *The New York Review of Books*, June 23, 2011, https://www.nybooks.com/articles/2011/06/23/mind-control-and-internet/.
341 Young Republicans seem to be a bit more conscious of the environment. Emily Goddard et al., "Explore 16 Years of U.S. Climate Opinions with the CCAM Explorer," *Yale Program on Climate Change Communication* (New Haven, CT), May 6, 2025, https://climatecommunication.yale.edu/publications/explore-16-years-of-u-s-climate-opinions-with-the-ccam-explore/.
342 Cass R. Sunstein, *Republic.Com 2.0* (Princeton University Press, 2007).
343 For a precise comparison of the similarities and differences: C Thi Nguyen, "Escape the Echo Chamber," *Aeon*, April 9, 2018, https://aeon.co/essays/why-its-as-hard-to-escape-an-echo-chamber-as-it-is-to-flee-a-cult.
344 Faruk Ateş, "A Brief History of Women in Computing | HackerNoon," August 9, 2017, https://hackernoon.com/a-brief-history-of-women-in-computing-e7253ac24306.
345 *Wikipedia*, "Sophia (robot)," September 29, 2025, https://en.wikipedia.org/w/index.php?title=Sophia_(robot)&oldid=1314142273.
346 Incidentally, we should not take the granting of citizenship too seriously: it was primarily an effective marketing gimmick.
347 Michael Anderson and Susan Leigh Anderson, eds., *Machine Ethics* (Cambridge University Press, 2011), https://doi.org/10.1017/CBO9780511978036.
348 Cf. Ray Kurzweil, *The Singularity Is Near: When Humans Transcend Biology* (Penguin, 2005)

349 Benedict Smith and Kieran Kelly, "A 14-Year-Old Boy Fell in Love with a Flirty AI Chatbot. He Shot Himself so They Could Die Together," The Telegraph, October 24, 2024, https://www.telegraph.co.uk/us/news/2024/10/24/teenage-boy-killed-himself-fall-love-ai-chatbot/.
350 Sherry Turkle, "Authenticity in the Age of Digital Companions," *Interaction Studies* 8, no. 3 (2007): 501–17, https://doi.org/10.1075/is.8.3.11tur.; Sherry Turkle et al., "Relational Artifacts with Children and Elders: The Complexities of Cybercompanionship," *Connection Science* 18, no. 4 (2006): 347–61, https://doi.org/10.1080/09540090600868912.
351 David Hume, *Enquiries Concerning Human Understanding and Concerning the Principles of Morals*, 3d ed., ed. L. A. Selby-Bigge and P. H. Nidditch (Clarendon Press, 1975), 95.
352 Ian Carter, "Positive and Negative Liberty," in *The Stanford Encyclopedia of Philosophy*, Spring 2022, ed. Edward N. Zalta (Metaphysics Research Lab, Stanford University, 2022), https://plato.stanford.edu/archives/spr2022/entries/liberty-positive-negative/.
353 Hegel was the first to state correctly the relation between freedom and necessity. To him, freedom is the appreciation of necessity. "Necessity is blind only insofar as it is not understood." Friedrich Engels, *Anti-Dühring. Herr Eugen Dühring's Revolution in Science* (Progress Publishers, 1947), 140, http://archive.org/details/antiduhring.
354 Henri Bergson, *Time and Free Will: An Essay on the Immediate Data of Consciousness*, 1st ed. (Routledge, 2014), 166–72.
355 Ryan J. Allred and John P. Crowley, "The 'Mere Presence' Hypothesis: Investigating the Non-verbal Effects of Cell-Phone Presence on Conversation Satisfaction," *Communication Studies* 68, no. 1 (2017): 22–36, https://doi.org/10.1080/10510974.2016.1241292.
356 Some people, like the offline-club, have started calling such boxes "phone hotels": "BBC World Service – People Fixing The World, Amsterdam's Phone-Free Cafe," BBC, August 21, 2024, https://www.bbc.co.uk/programmes/p0jkqfmr.
357 Carter, "Positive and Negative Liberty."
358 See, for example, *Drones Have Become the Dominant Killers in the Ukraine War – National Security News*, April 9, 2025, https://nationalsecuritynews.com/2025/04/drones-have-become-the-dominant-killers-in-the-ukraine-war/.
359 "Artificial Muscles Set to Create Robots with Super Powers," *The National*, November 29, 2017, https://www.thenationalnews.com/business/technology/artificial-muscles-set-to-create-robots-with-super-powers-1.679896.
360 Bryan Clark, "Boston Dynamics' New Robot Can Jump 4 Ft, Lift 100 Lbs, and Skate on 2 Wheels," *TNW* | Insider, February 28, 2017, https://thenextweb.com/news/boston-dynamics-new-robot-can-jump-4-ft-lift-100-lbs-and-skate-on-2-wheels.
361 "Robotic Cheetah 'Breaks Speed Record for Legged Robots,'" Technology, *BBC News*, March 6, 2012, https://www.bbc.com/news/technology-17269535.
362 Wikipedia, "List of mergers and acquisitions by Alphabet," October 6, 2025, https://en.wikipedia.org/w/index.php?title=List_of_mergers_and_acquisitions_by_Alphabet&oldid=1315404868.
363 Mark Berniker, "Google Founders' Private Air Terminal Cleared for Takeoff," CNBC, January 16, 2014, https://www.cnbc.com/2014/01/16/google-founders-ok-to-build-private-air-terminal-at-san-jose.html.
364 Miguel Helft, "A New (Fighter) Jet for Google's Founders?" Bits Blog, October 23, 2008, https://archive.nytimes.com/bits.blogs.nytimes.com/2008/10/23/a-new-fighter-jet-for-googles-founders/.
365 Wikipedia, "List of countries by GDP (nominal)," December 4, 2025, https://en.wikipedia.org/w/index.php?title=List_of_countries_by_GDP_(nominal)&oldid=1325623456.

366 Daisuke Wakabayashi and Scott Shane, "Google Will Not Renew Pentagon Contract That Upset Employees," Technology, *The New York Times*, June 1, 2018, https://www.nytimes.com/2018/06/01/technology/google-pentagon-project-maven.html.
367 https://brill.com/view/journals/ihls/16/2/article-p336_003.xml?language=en&srsltid=AfmBOoruUF14BLMTS5dB9Os_e0VdSs8IPzv8PU3Uxf1nTTpA4n0LMOJOhttps://www.setav.org/en/gaza-as-a-testing-ground-israels-ai-warfar
368 See, for example, Cora Morris, "Death on Delivery: New Report Examines Civilian Harm from Intensifying Drone Warfare across Africa," *Drone Wars UK*, March 10, 2025, https://dronewars.net/2025/03/10/death-on-delivery-new-report-examines-civilian-harm-from-intensifying-drone-warfare-across-africa/.
369 P. W. Singer and Emerson T. Brooking, *Likewar: The Weaponization of Social Media* (Houghton Mifflin Harcourt, 2018).
370 S. Spiekermann and F. Pallas, "Technology Paternalism – Wider Implications of Ubiquitous Computing," *Poiesis & Praxis* 4, no. 1 (2006): 6–18, https://doi.org/10.1007/s10202-005-0010-3.
371 Amartya Sen, "Capability and Well-Being," in *The Quality of Life*, ed. Martha Nussbaum and Amartya Sen (Oxford University Press, 1993), https://doi.org/10.1093/0198287976.003.0003.
372 "STADA Health Report 2025," STADA, 2025, https://www.stada.com/media/health-reports/stada-health-report-2025.
373 On any given working day: Karl Thompson, "What Percentage of Your Life Will You Spend at Work? – ReviseSociology," August 16, 2016, https://revisesociology.com/2016/08/16/percentage-life-work/.
374 *State of the Global Workplace Report* (Gallup, 2025), 6, https://www.gallup.com/workplace/349484/state-of-the-global-workplace.aspx.
375 April Dembosky and Emily Steel, "Health Apps Run into Privacy Snags," Financial Times, *Financial Times*, September 1, 2013, https://www.ft.com/content/b709cf4a-12dd-11e3-a05e-00144feabdc0.
376 A strike by those affected ultimately led to the abolition of this system. See Gabrielle, "How Employers Have Gamified Work for Maximum Profit | Aeon Essays."
377 Richard H. Thaler and Cass R. Sunstein, *Nudge: Improving Decisions about Health, Wealth, and Happiness* (Yale University Press, 2008).
378 The questions may sound critical. In reality, families today often immerse themselves in virtual worlds and play together. This type of shared play and experience in virtual worlds can be very positive for family cohesion; without having researched it, it may lead to parents and children getting to know each other better and recognizing each other's potential. However, the same game can also be harmful if children immerse themselves in these worlds in isolation, spending weeks and months of their lives there.
379 Wikipedia, "Junzi," October 6, 2025, https://en.wikipedia.org/w/index.php?title=Junzi&oldid=1315467003.
380 It might be worth considering linking psychological motive research to the philosophical teachings of telos. Ultimately, psychologists have empirically researched a great many motives; however, they do not yet have a complete answer as to how motives develop and why they vary so much among people.
381 Johann Wolfgang von Goethe. *Materialien zur Geschichte der Farbenlehre*. Translated in Crispian Villeneuve, *Rudolf Steiner: The British Connection : Elements from His Early Life and Cultural Development* (Temple Lodge Publishing, 2011), 367.
382 Kyle Chayka, "Are You Experiencing Posting Ennui?," Infinite Scroll, *The New Yorker*, July 9, 2025, https://www.newyorker.com/culture/infinite-scroll/are-you-experiencing-posting-ennui.

383 Kostadin Kushlev and Elizabeth W. Dunn, "Checking Email Less Frequently Reduces Stress," *Computers in Human Behavior* 43 (February 2015): 220–28, https://doi.org/10.1016/j.chb.2014.11.005.
384 On any given working day: Karl Thompson, "What Percentage of Your Life Will You Spend at Work? – ReviseSociology," August 16, 2016, https://revisesociology.com/2016/08/16/percentage life-work/.
385 Eric J. Horvitz et al., Notification platform architecture, World Intellectual Property Organization Patent WO2001069387A8, filed March 16, 2001, and issued December 6, 2001, https://patents.google.com/patent/WO2001069387A8/en.; Eric Horvitz et al., "Models of Attention in Computing and Communication: From Principles to Applications," *Commun. ACM* 46, no. 3 (2003): 52–59, https://doi.org/10.1145/636772.636798.

Index

 | https://doi.org/10.1515/9783111677736-012

www.ingramcontent.com/pod-product-compliance
Lightning Source LLC
Chambersburg PA
CBHW070338290826
50605CB00036B/527
* 9 7 8 3 1 1 1 6 7 7 6 3 7 *